ENGINEERING
ARTIFICIAL INTELLIGENCE

ENGINEERING
ARTIFICIAL INTELLIGENCE

Editor

I. M. Makarov

USSR Academy of Sciences
Moscow

English Edition Editor

E. I. Rivin

Wayne State University
Detroit, Michigan

⬤HEMISPHERE PUBLISHING CORPORATION

A member of the Taylor & Francis Group

New York   Washington   Philadelphia   London

Authors: **V. M. Nazaretov** and **D. P. Kim.**

ENGINEERING ARTIFICIAL INTELLIGENCE

Originally published as Tekhnicheskaya imitatsiya intellekta by Vysshaya Shkola, Moscow, 1986.

Translated by P. N. Budzilovich.

1 2 3 4 5 6 7 8 9 0   B R B R   9 8 7 6 5 4 3 2 1 0

This book was set by Allen Computype.
Cover design by Reneé E. Winfield.
Printing and binding by Braun-Brumfield, Inc.

A CIP catalog record for this book is available from the British Library.

Library of Congress Cataloging-in-Publication Data

Nazaretov, V. M.
   [Tekhnicheskaia imitatsiia intellekta. English]
   Engineering artificial intelligence / [authors, V.M. Nazaretov and D.P. Kim ; translated by P.N. Budzilovich] ; editor, I.M. Makarov, E.I. Rivin.
     p.   cm.
   Translation of: Tekhnicheskaia imitatsiia intellekta.
   Title on verso t.p.: Tekhnicheskaya imitatsiya intellekta.
   1. Robotics. 2. Automation. 3. Artificial intelligence.
I. Kim, D. P. II. Makarov, Igor Mikhaĭlovich. III. Rivin, Eugene I. IV. Title. V. Title: Tekhnicheskaya imitatsiya intellekta.
TJ211.N3913 1990
629.8'92—dc20
ISBN 0-89116-963-6

90-4379
CIP

D
6 29. 8 9 2
E N G

CONTENTS

INTRODUCTION

The development of highly automated labor-saving factories presupposes the automation of not only manual, but also intellectual, human tasks. The automation of intellectual activity demanded a solution to a number of new problems that had not arisen previously in the theory of automatic control. Problems include the description and representation on a computer of a complex environment, automated planning and performance of a variety of various operations by mechanisms aimed at achieving a specified goal, the development of communications between humans and computers using a natural language, and a number of other problems.

The scientific discipline encompassing efforts aimed at the solution of these problems is referred to as artificial intelligence. It borders on a variety of disciplines, including mathematical logic, information theory, control theory, computing devices, programming, etc. The scope of research in the area of artificial intelligence is characterized not only by a diversity of themes, but also by its range—from the strictly abstract, aimed at the development of general principles or a theory of artificial intelligence, to engineering, aimed at the development of hardware and software for solving intellectual problems.

This book does not set out to embrace the entire area of artificial intelligence. The main purpose of a textbook is to present the fundamentals of such parts of this discipline which, to some degree, survive the test of time and may be used to design systems for a specific purpose, e.g., in robotics and automated manufacturing.

The book is based on lectures for the course "Engineering Simulation of Intelligence," which was taught by the authors at the Moscow Institute of Radio, Electronics, and Automation, to students specializing in robotic systems.

ARTIFICIAL INTELLIGENCE IN ROBOTICS AND FLEXIBLE MANUFACTURING SYSTEMS (FMS)

The expression "artificial intelligence" is used in a dual sense: as an engineering informational model of natural (human) intelligence and as a scientific/engineering discipline or a scientific school that deals with problems of human intelligence simulation.

Artificial intelligence today comprises an independent, rapidly developing scientific discipline. Major research conducted in the area of artificial intelligence may be reduced to the following for directions [1]:

1. Representation of knowledge objects and manipulations with them. This includes the development of specialized models and languages for representing knowledge in a computer, as well as the software and hardware for their transformations (enrichment, logical processing, etc.). Research is being done in the area of the development of specialized logic systems permitting the enrichment and generalization of knowledge stored in a computer.

2. Planning of rational behavior. This includes research aimed at the development of methods for formulating goals and solutions of problems of planning operations of an automated device operating in a complex environment.

3. Man-computer communications. This area includes the problems of developing languages for effective interaction between a lay user (non-programmer). Research is being done in the area of syntax and semantics of natural languages, on methods for storing knowledge about a language in the computer memory, and on the development of special processors for implementing the transfer of text information into the internal machine notation.

4. Object recognition and learning. Research here involves perception of visual, audio, and other types of information, its processing methods, formulation of responses to the environment, and methods of adaptation of artificial systems to an environment by teaching.

1

1.1 THE CONCEPT OF ARTIFICIAL INTELLIGENCE SYSTEMS

Artificial intelligence systems or "intelligent systems" are those that perform functions that are considered intelligent.

The intelligent activity of man is connected to the search for solutions (actions, natural phenomena) in novel, nonstandard situations. Accordingly, a problem is termed intelligent if a plan or a precise (algorithmic) method of its solution is not known *a priori*. Here the problem and its solution are understood in a broad sense. Problem solution is any activity (by human or machine) aimed at the development of plans and actions necessary for the achievement of some specific goal, derivation of new correlations, formulation (by a machine) of sentences in a natural or familiar language, etc.

Any intellectual activity is based on knowledge of some object region within which the problems are posed and solved. The "object" or "problem region" generally refers to a set of interrelated information necessary and sufficient for the solution of a given problem or a specific set of problems. Knowledge concerning an object region includes descriptions of objects, events, and facts, as well as their interrelations.

The role of knowledge in an intellectual activity determines the characteristic feature of intelligent systems, that is, the inclusion into them of a block of knowledge representation (BR3). This block is connected with the "external" world by two transforming blocks (BR1 and BR2) which transform the knowledge about an object region (including requests) from an external (ER) into an internal (IR) representation — (BR1) and, conversely, from an internal into an external representation, (BR2), which can be "understood" by the user (see Fig. 1.1a). In general, an information model of the block of knowledge representation consists of an interpreting block (BI), a teaching block (BT), a solution output block (BO), and the intelligent database (IDB) (see Fig. 1.1b). The intelligent database, in turn, is divided into a knowledge database (KDB) and data database (DDB).

The knowledge database contains information that reflects laws and interrelationships within a given object region and permits the forecast and derivation of new facts that have not been reflected in the database.

The database includes factual, quantitative data that characterize a given object region.

In general form, knowledge in a computer is represented by some symbol (semiotic) system. The notion of a "symbol" is directly related to the notions "denotation" and "concept (attribute)." The denotation is the object denoted by a given symbol, while the attribute is a property of the denotation. The important notions in semiotic systems are extension and intention. The extension of a symbol defines a specific class of all of its permissible denotations. The intention of a symbol defines the content of a notion associated with it. Correspondingly, there are intentional and extentional types of knowledge. The intentional knowledge describes abstract objects, events, and relations, such as, VENDOR, USER, TRANSPORTATION. The extentional knowledge represents data that

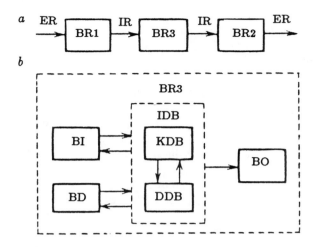

Figure 1.1   Knowledge transformation diagram (*a*) and the information model of the knowledge representation block (*b*).

characterizes specific objects, their states, and parameter values at some specific instants of time. For example, an extension of the notion VENDOR might be COMPANY A, an extension of the notion USER might be ORGANIZATION B, and an extension of the notion TRANSPORTATION might be AUTOMOBILE.

In a semiotic system, three aspects are isolated: syntatic, semantic, and pragmatic. The syntax describes the internal arrangement of a symbol system, i.e., rules of construction and transformation of complex symbolic expressions. For the natural languages, as is well known, syntax defines the correct construction of sentences and of the related text. Semantics defines relations among the symbols and their concepts, i.e., it assigns the sense and meaning to specific symbols. Pragmatism defines a symbol from the point of view of its usage or of a subject utilizing a given symbolic system.

In accordance with the above three aspects of semiotic systems, we can define three types of knowledge: syntactic, semantic, and pragmatic. The syntactic knowledge characterizes the syntactic structure of an object or an event being described that does not depend on the sense or meaning of notions used in the process. The semantic knowledge contains information that is directly related to the sense or meaning of objects or events being described. The pragmatic knowledge describes objects and events from the point of view of the problem being solved; for instance, by taking into account specific criteria applicable to the given problem.

Corresponding to the three types of knowledge, there are three types of models for their representation: syntactic, semantic, and pragmatic. The presence of semantic or pragmatic models most significantly differentiates intelligent systems from all other various hardware-software systems that can be implemented using modern computers.

In the manufacturing industry, there are two main areas where the ideas and methods of artificial intelligence can be applied: robotic systems and management-economic and design systems. The first is associated with the development of third generation industrial robots, i.e., robots with the elements of artificial intelligence. The second is associated with the development of management-manufacturing systems on the basis of decision-making hardware and software in such endeavors as product design, technology development, planning, management, quality control, etc. In this area we have three types of intelligent systems: question-answer (dialogue) systems permitting one to interact with databases using a limited natural language; computational-logical systems permitting one to automatically select a problem solution method and to formulate the problem on the basis of the problem description in terms of the user's domain of expertise; and expert systems, or systems-consultants, permitting the generalization, storing, and use of knowledge and experience accumulated by highly qualified specialists in a given subject area.

1.2 STRUCTURE AND FUNCTIONS OF AN INTELLIGENT ROBOT

The main characteristic of an intelligent robot is its purposeful behavior in a complex, poorly-organized environment. From the point of view of artificial intelligence, the purposeful behavior may be organized by means of the transformation of knowledge concerning the current state of the environment, obtained by means of sensors, into a sequence of operations aimed at the achievement of a predetermined goal. Such a transformation must be based on *a priori* knowledge about the environment and the methods of its transformation. This means that central units within an intelligent robot are a system of knowledge representation and a developed system of operation planning. Another important characteristic of an intelligent robot is that all data acquisition and transformation processes must occur in real time.

A generalized functional block diagram of an intelligent robot is shown in Fig. 1.2. It can be broken down into three separate interrelated systems: perception, knowledge representation, planning and actuation. Let us consider each of these systems.

It is convenient to begin our analysis with the system of knowledge representation, since the other two systems, to a large degree, depend upon it. There are three aspects connected with knowledge in intelligent robots: knowledge representation; storage of knowledge; and its updating and utilization in solving problems. Knowledge representation (e.g., the mode in which it is expressed) should be chosen by taking into account the specific types of problems that the robot is intended to solve. Representation modes are treated in Chapter 2. Here the knowledge representation system will be viewed as an aggregate of four blocks: abstract

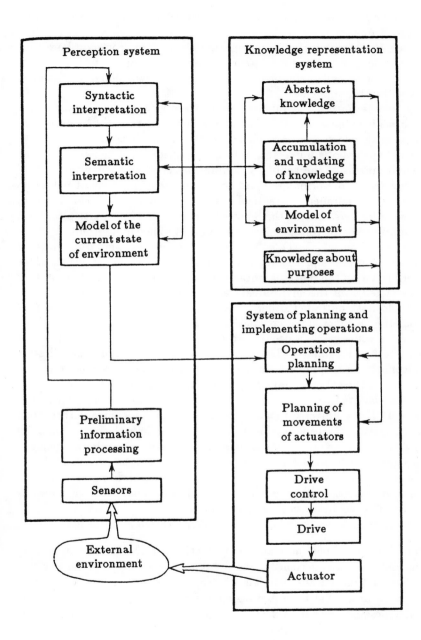

Figure 1.2 Functional block diagram of an intelligent robot.

knowledge, knowledge of the purposes, model of the robot's universe, storage and updating of knowledge.

The abstract knowledge includes information about some common laws that are valid both in the external and the internal universe environment of the robot, which are usually considered as conditionally constant. They include, for instance, physical laws within the universe.

Knowledge of the purposes is information concerning the global goals which the robot must achieve in its operation, as well as the methods of their decomposition into local goals which can be realized during intermediate steps.

The model of the robot's universe is a formal description of knowledge of the operational environment of the robot. This information is of an *a priori* nature in the sense that it was preformulated and communicated to the robot before its operation began. In a number of practical cases it is impossible to construct an initial model of the universe having the required accuracy and information content. This is particularly true for robots operating in poorly-known environments.

Current information which the robot receives during its operation may be used to increase accuracy and to broaden its knowledge of the universe. This function is performed by the block of storage and updating of knowledge. Within the block, new facts about the environment are accumulated and analyzed to ensure that they do not contradict other facts stored in the system of knowledge representation. If a new fact is not contradictory, it may be included in the model of the universe. In some cases, a decision to include a new fact into the model is preceded by a check of its validity. This entails either awaiting more data directly or indirectly confirming or refuting the fact being questioned, or performing an active experiment in the external universe to confirm or refute the fact. An analysis of a fact which contradicts facts already included in the model is performed in a similar manner. If the universe model is required to be noncontradictory, then an admission of a new fact, recognized as a valid one, demands the exclusion of all noncomplying facts. However, the noncontradiction is not a mandatory requirement imposed upon systems of knowledge representation. The "tolerance" of a system to contradictions enables its stable operation in a poorly-known environment and, to some extent, corresponds to the ability of a man to make decisions in spite of contradictions between some basic facts.

Thus, specific requirements to knowledge representation systems are:

- "tolerance" to incompleteness and contradictions. Imperfections in sensing systems and limited experience make the robot's knowledge of the universe incomplete and inaccurate. The knowledge representation system must be capable of ensuring that the robot does not stop operating when it uncovers incompleteness or inaccuracies in its knowledge. Only a reduction in the efficiency of its operation is permitted.

- the ability to evaluate new information. This is the ability to check the conformity of new information with the previously obtained data and to make a decision concerning its validity.

- the ability to learn and update knowledge. In conjunction with the ability to evaluate new information, learning ability should expand knowledge and increase its reliability.

An intelligent robot communicates with the external universe by means of its perception (sensing) system. The eventual goal of this system is the construction of a model that represents the current state of the universe. Sometimes it is said that the perception system formulates the situational knowledge of a robot, i.e., its knowledge of the current situation. As in second-generation robots, in the intelligent robots the prime source of information about the universe are sensors that include tactile, positional, force, vision, audio, and others. Sensor information is processed and presented in a convenient form for subsequent manipulations. This part of a perception system of an intelligent robot does not significantly differ from similar systems of second-generation robots equipped with perceptive systems. Subsequent transformation of information related to its syntactic and semantic interpretation is characteristic only of intelligent robots. During the syntactic interpretation, the representation of the universe is formulated using certain internal language, while the meaning of the perceived events and objects remains hidden. The semantic interpretation is related to the discovery of the meaning of the perceived information. The final procedure is the construction of a model of the current state of the universe. Let us note in passing that the syntactic and semantic interpretations, as well as the synthesis of the model of the current state of the universe, are impossible without the knowledge of the robot's universe. For this reason, the realization of these procedures is performed as an active two-way interaction with the knowledge representation system.

The main purpose of the planning and actuation system is to formulate and implement a program of operations affecting the external universe and leading to the achievement of the desired goal. Planning of activities of an intelligent robot is usually implemented as a problem solution process. In so doing, the problem is viewed in its broadest sense as the difference between the current and desired states of the universe. The plan or the problem is a sequence of operations that transform some current state of the universe into a desired state. Thus, for an assembly robot, the plan will be a sequence of operations aimed at changing the positions of parts from their current state into the desired state corresponding to the assembled product. Here the term "operation" means such operations as "join part A with part B," "screw bolt into...," etc. Obviously, to carry out these operations, they should be broken down into basic movements that can be performed by the manipulator. This function is performed by the movement planning block of the actuator that formulates the program of movements to be implemented by the drive control.

Intelligent robots are frequently referred to as "integrated." The term "integrated robot" was first used to describe a robotic system equipped with machine vision, tactile sensors, and a movement capability. At the present time, the term "integrated robot" refers to a robotic system that contains five groups of functionally completed systems [4]:

- Group V includes perceptive systems of visual, audio, tactile, and other types of information about the external environment;

- Group M includes systems for acting upon external objects (manipulators, manufacturing systems, etc.);

- Group T includes robot mobility systems;

- Group P includes systems for planning operations and problem solving;

- Group R includes communication systems between the robot and operator and/or other robots.

Any specific robot may be formed by combining all or some of the above systems. This permits the expression of the degree of robot integration by a number of systems which comprise it and permits the definition of classes of robots based on the degree of integration and types of their component systems.

Integration of order *1* apparently consists of five classes. Class M may include robot-manipulators of the first generation, Class T may include the simplest transportation robots. The remaining three classes are vacant, since the systems belonging to these groups cannot be viewed as independently robotic systems.

Integration of order *2* consists of 10 classes. Class VM may include sensualized industrial robots, Class VT may include sensualized transportation robots, and Class MT may include mobile robot-manipulators.

Integration of order *3* consists of 10 classes. For instance, Class VMP may include sensualized robot-manipulators of the third generation equipped with sensors and an activity planning system, Class VTP may include third-generation transportation robots.

Examples of integration of the *4*th order are Class VMTP robots, that is, sensualized mobile robot-manipulators with operations planning systems, and Class MTPR robots, that is, mobile robot-manipulators with speech-recognition input command interfaces.

Integration of order *5* consists, apparently, of only one class: VMTPR, i.e., a class of robots that includes all the above-mentioned systems.

Intelligent robots also include integrated autonomous robots with integration orders *3*, *4*, and *5*. Here the term "autonomous" refers to the ability to independently (or relatively independently) receive information concerning the environment and plan and to implement operations required to achieve a specified goal. The autonomous property presupposes that the robot contains systems of groups V and P.

1.3 INTELLIGENT SYSTEMS IN HIGHLY-AUTOMATED MANUFACTURING

The development of modern manufacturing has meant the continuous growth of its complexity, expressed in the broadening of the diversity of technologies that utilize new physical phenomena and engineering solutions, as well as in the acceleration of the changes and complexity of manufactured goods. The knowledge required for an effective design and utilization of manufacturing is continuously increasing, thus significantly increasing the intellectual burden upon designers and manufacturing engineers. For instance, the utilization of new types of automation based on electronic computing technology drastically increases the intellectual labor component of manufacturing personnel from manager down to laborer. Traditionally, problems of this kind were solved by increasing the specialization of individual workers, which invariably resulted in personnel increases. This approach contradicts the tendency to reduce the number of designers and manufacturing personnel and to develop labor-saving technologies.

This problem is being attacked from two directions. On the one hand, the education level of engineering, design, and manufacturing personnel is being increased; and on the other, the degree of intelligence of automated design and control systems is also being increased.

Design. The lack of engineering methods for calculating economic efficiency and functional characteristics of a Flexible Manufacturing System (FMS) complicates the design process and increases the role of knowledge and experience accrued while designing such systems, as well as of new methods for their analysis and synthesis by means of mathematical modeling. The basis for generalizing and disseminating knowledge obtained during research and experiments in developing FMS are the expert systems that may be employed, for instance, during a predesign (the feasibility study). The purpose of an expert system during this phase is to determine the economical and technological feasibility of an FMS on the basis of such parameters as productivity, type of product, mass-producibility, proposed manufacturing technologies, and others, as well as to indicate which structure classes are the best for each specific case.

One method which is widely used in automated FMS design is mathematical modeling by means of a computer (simulation, for instance). Using this method, one can conduct controlled experiments with the system model. Model development and the associated experiments require good mathematical and programming training, as well as broad engineering knowledge of FMS. This complicates the process of interactive design where the designer directly interacts with a computer in a dialogue mode. In traditional systems, "middle men" are used between the user and a computer, usually mathematicians and programmers. In intelligent systems, a direct interaction between the user and computer is provided by an intelligent interface, that is, hardware and software that permit one to conduct a dialogue in the designer's language.

Manufacturing management. One can identify at least three areas where intelligent systems may be used in manufacturing management.

1. Automated solution of problems in planning and control on the basis of computational-logical and expert systems. The necessity of using computational-logical systems stems from the following. At this time, there exists

a substantial library of mathematical methods of planning and operational management. Its utilization requires the use of mathematicians and programmers. This, however, is impossible to realize in practice at highly-automated manufacturing facilities with limited staff. As was pointed out earlier, computational-logical systems select the problem solution method and formulate a program for describing the problem and quantitative data inputted into the computer using a limited professional language of the user.

The use of expert systems stems from the fact that planning and control functions that are expected to be performed by computers are continuously broadening, which inevitably leads to the necessity to formalize knowledge and experience of appropriate specialists. The most convenient — and often the only possible — method for such formalization is the representation of knowledge in the form of semiotic models and logic derivation rules. An expert system, constructed on the basis of such models and rules, unlike a computational-logical system, does not solve a problem. Instead, it "consults" the planner or an engineer, and diagnoses complex conditions, i.e., it acts as a system-advisor.

2. Automated programming using an intelligent interface. Automatic programming systems with elements of artificial intelligence permit a nonprogrammer (an operator, for instance) to develop programs for performing industrial operations. Here one frequently uses programming systems that are based on machine drawings, since the language of graphic images (drawings, for example) is well known to manufacturing personnel. Programming may also be realized on the basis of voice commands. To receive voice, the system is equipped with a speech analyzer.

There is some experience in outfitting industrial equipment with speech synthesizers to permit the arrangement of a voice dialogue of an operator and a computer controlling the equipment. Speech synthesizers may be used to signal the operator by voice about equipment malfunctions or special conditions of the equipment, and even for training an operator to work with the equipment*.

3. Automation of technology using machine vision systems. There is a wide class of industrial processes that may be automated using visual information. Typically, this includes assembly processes.

There are two types of automated systems that use machine vision. In systems of the first type, visual information is used in a nonorganized environment to identify objects and to determine their position in space; in systems of the second type, information is used for controlling a process. The environment is assumed to be sufficiently organized, so that the process may be performed in accordance with a predetermined program. The realization of this program does not require any visual information. The

*See Vol. 7 of this series.

system of machine vision in this case is used to detect deviations from normal functioning. Typical examples include visual quality control of parts or blanks, correct sequence in performing a manufacturing operation, etc.

TWO
REPRESENTATION OF KNOWLEDGE

The term "knowledge" as regards artificial intelligence is just as significant as the term "data" in programming. The transfer from data to knowledge is the logical consequence of the development and increasing complexity of information structures processed by computers.

2.1 GENERAL INFORMATION ON MODELS FOR KNOWLEDGE REPRESENTATION

Let us consider aspects of knowledge which differentiate it from data [1].

1. Interpretability. Data stored in a computer may be interpreted in a meaningful manner only by means of an appropriate program. Without a program, data contains no meaningful information. Knowledge differs from data in that it always offers the possibility of meaningful interpretation.

2. The availability of classifying relationships. In spite of the variety of data storage forms, not one of them permits a compact description of all inter-relations among different types of data. For instance, in working with data that characterizes both the individual elements of a set and the entire set, one has to repeatedly describe the same information, common to both the elements and the entire set. On the contrary, it is possible to establish among individual units of knowledge such relationships as "element-set," "type-subtype," "situation-subsituation," that reflect the nature of their relationship. This permits the separate writing and storing of information that is common for all elements of a set. If required, this information may be automatically transferred to the description of any element of the set. This transfer process is referred to as information "inheriting."

3. The availability of situational correlations. These correlations determine the situational compatibility of separate events or facts stored or inputted into the memory, as well as such relations as simultaneousness, location in the

same region of space, being in a state of mechanical interaction, etc. Situational correlations help construct procedures for knowledge analyses with respect to compatibility, contradiction, etc. which are difficult to realize by storing traditional data blocks.

The advent of knowledge as an object of information to be processed on a computer caused the transition from data bases to knowledge bases. Knowledge base management systems (KBMS) have evolved from database management systems (DBMS). KBMS offer the user more powerful servicing procedures than those possible with DBMS. For instance, the user can work not only with those information structures that are realized in a knowledge base, but can also create his own. KBMS automatically establishes communications among user structures and structures stored in the knowledge base. Let us note that a knowledge base neither rejects nor replaces data bases. As was shown in §1.1, a knowledge base and a database are viewed as different levels of representation of information stored in an intelligent information bank.

The central question in creating a knowledge base is the selection of a method of representation (e.g., description) of knowledge and communication with the user. In combination, a model of knowledge representation and the associated procedures forms a knowledge representation system (KRS).

Forms of knowledge representation may be divided into declarative and procedural. The declarative representations do not contain, in an explicit form, descriptions of procedures that should be performed. The declarative knowledge is, as a rule, a set of statements that do not depend on where they are used. Modeling an object region in this form requires a full description of its state which is syntactic in nature. The derivation and search for solutions are mostly based on search procedures in the state space. The development of such procedures hinges on the specifics of the given object region, i.e., by taking into the account its semantics (see §3.2). Consequently, in a declarative representation, semantic and syntactic knowledge are to some degree isolated from one another. This results in a greater universality and commonality in this form of representation.

The procedural knowledge contains the description of certain procedures in an explicit form. Here a current state is represented in the form of a set of specialized procedures for processing a specific sector of the knowledge base. This permits the storing of simply a certain initial state and a procedure for generating the necessary descriptions of states on the basis of this initial state. The storing of the descriptions of all possible states required for developing a conclusion or solution is not necessary. In procedural representation, semantics is introduced into the description of knowledge base elements. This permits an increase in the efficiency of the solution search because of the possibility to utilize more involved computational techniques and because of the exclusion of the necessity to process full descriptions. Procedural representations, as a rule, are implemented by using specialized languages such as PLANNER, CONNIVER, and others. Knowledge representation in this form ensures faster solution search compared to the declarative representation, but it does not offer the same capabilities as far as storage and updating of knowledge are concerned.

The division of representation forms into declarative and procedural is conditional, since specific models utilize both forms of representation in varying degrees. Logical and network models of knowledge representation are most widely used.

At the basis of logical models is the concept of a formal system defined by the quad $\mathbf{M} = (\mathbf{T}, \mathbf{P}, \mathbf{A}, \mathbf{F})$, where $\mathbf{T}$ is a set of base elements; $\mathbf{P}$ is a set of syntactic rules that permit the construction of syntactically correct expressions from $\mathbf{T}$; $\mathbf{A}$ is a set of *a priori* true expressions; $\mathbf{F}$ is the semantic derivation rules that permit a broadening of a set of axioms at the expense of other expressions. The utilization of logic of different types in the development of syntactic and semantic rules results in logic models of different types. Models based upon calculus of predicates are used most widely. Development of the predicate systems particularly accelerated after the advent of powerful derivation procedures of the resolution-type method (see §3.4).

Network models of knowledge representation, unlike logical models, offer increased capabilities for describing complex structures that characterize knowledge. This is achieved by identifying and including into the model in an explicit form all relationships that form an informational structure with the description of its semantics. The basis of these models is a network. Vertices are identified with certain concepts (attributes), while the arcs are identified with relationships between them. Vertices may have their own internal structure. Network models of two types are widely used: semantic networks and frames.

The representation of linguistic knowledge is directly related to the problem of developing intelligent systems that provide an interface (two-way communication) with the user employing a natural or a near-natural language. The realization of such an interface presupposes the availability of models of formal description of the language and text-processing algorithms, as well as software and/or hardware supporting the utilization of these models and algorithms. The combination of the above-mentioned components determines a linguistic knowledge representation system (LKRS) commonly divided into model-algorithmic and software-hardware parts. The theory of formal language is the basis of the model-algorithmic part. The software-hardware part is realized by a language processor (L-processor). The purpose of L-processors is direct and reverse translation of natural-language texts into a formal machine language of a given computer. In an L-processor, there are declarative and procedural parts. The first contains the description of the L-processor dictionaries; the second contains algorithms of analysis and synthesis of natural-language texts.

2.2   LOGICAL MODELS OF KNOWLEDGE REPRESENTATION

To describe the external universe and to search for solutions, the art of artificial intellect freely uses the techniques and language of calculus of predicates. The calculus (or algebra) of predicates is the development of the calculus of statements and includes it in its entirety as a component part. Therefore, we will introduce the calculus of predicates by reviewing the calculus of statements.

Calculus of statements. In mathematical logic, the term "statement" refers to a proposal that is either true or false. Let us denote statments by capital Latin letters A, B, C, ... with or without subscripts and refer to them as propositional. Since the statements (propositions) may be true or false, propositional symbols may take on truthfulness values T (true) and F (false). Sometimes, as in the theory of finite automation, 1 and 0 are used instead of true and false, respectively.

More complex propositions may be obtained by employing truthfulness-functional correlations F $(A$, B, $...)$, which are determined by truthfulness (truth) values A, B, ... The functions F $(A$, B, $...)$ are usually given in the form of tables that are referred to as "truth tables." A "truth function" (or a "function of truth algebra") with n arguments is any function with n truth arguments that take on truth values T or F (1 or 0). Truth arguments are those that assume truth values T or F (1 or 0).

Logical operations. On the basis of given propositions, one may form complex (composite) propositions by means of logical operations. The simplest composite propositions are as follows:

1) $\overline{A}$ (other notations: $\neg A$, $\sim A$, $-A$) is the negation of A or complement of A; it is read as "not A." A proposition is true (equals T) if proposition A is false (equals F); $\overline{A}$ is false if A is true;

2) $A \wedge B$ (other notation: $A \cdot B$, $A \& B$) is the conjunction or logical multiplication of A and B; it is read as "A and B." This proposition is true in that (and only in that) case when both A and B are true; A and B are referred to as conjunctive terms;

3) $A \vee B$ is the disjunction or logical addition of A and B; it is read as "A or B." This proposition is true in that (and only in that) case when at least one of propositions A and B is true; A and B are referred to as conjunctive terms;

4) $A \rightarrow B$ (another notation: $A \supset B$) is the implication or following of B from A; it is also read as "if A, then B." This proposition is false in that (and only in that) case when A is true and B is false;

5) $A \equiv B$ is the equivalence; it is read as "A then and only then, when B." This proposition is true then and only then when A and B have the same truth value.

Symbols of logic operations $-$, $\wedge$, $\vee$, $\rightarrow$, $\equiv$ are also referred to as propositional signs.

Below is a truth table that defines the elementary composite propositions introduced above (Table 2.1).

Table 2.1 (F — false, T — true)

A	B	$\overline{A}$	$A \wedge B$	$A \vee B$	$A \to B$	$A \equiv B$
F	F	T	F	F	T	T
F	T	T	F	T	T	F
T	F	F	F	T	F	F
T	T	F	T	T	T	T

Propositional forms. Each complex proposition made up from some input propositions by means of logical operations is known as a "logical algebra formula," or as a "propositional form."

More precisely, a propositional form (or logical algebra formula) is defined as follows (a recursive definition):

1) All propositional symbols (characters) are propositional forms;
2) If H and G are propositional forms, then $\overline{H}$, $(H \wedge G)$, $(H \vee G)$, $(H \to G)$, and $(H \equiv G)$ are propositional forms;
3) No other expressions are propositional forms. For brevity, propositional forms will also be referred to as "forms" or "formulae."

Each form constitutes a truth function. Different forms may constitute the same truth function. Two different forms representing the same truth function are referred to as equivalent. If H and G are equivalent forms, then we will write $H = G$. A form which is true under any truth values of its arguments is known as "tautology." Examples of tautology are $(A \vee \overline{A})$, that is, the "law of the excluded third," $(A \wedge \overline{A})$, $(A \equiv \overline{A})$. A form which is false (i.e., assumes the value F) under all possible truth values of its arguments is referred to as a "contradiction." Examples of contradictions are $(A \equiv \overline{A})$, $(A \wedge \overline{A})$.

The logic operations reviewed above are not independent. Some of them may be expressed by means of others. For instance, logic operations of implication and equivalence may be expressed by means of others. Indeed, it is easy to verify by means of truth tables that the form $(A \to B)$ is equivalent to the form $(\overline{A} \vee B)$, and that the form $(A \equiv B)$ is equivalent to the form $(\overline{A} \vee B) \wedge (\overline{B} \vee A)$

$$(A \to B) = (\overline{A} \vee B)$$

$$(A \equiv B) = ((\overline{A} \vee B) \wedge (A \vee \overline{B}))$$

Therefore, in what follows, we will pay most attention to properties of three logical operations: negation, conjunction, and disjunction.

Main properties of logical operations. For simplicity, let us use **1** and **0** for truth values (instead of T and F, respectively). Using a truth table to represent a truth function, it is not difficult to establish the following properties of logical operations.

1. Conjunction and disjunction operations have the property of being "associative"

$$((A \wedge B) \wedge C) = (A \wedge (B \wedge C))$$

$$((A \vee B) \vee C) = (A \vee (B \vee C))$$

Therefore, proportional forms containing only conjunction signs or only disjunction signs may be written without internal parentheses, while the external parentheses may be omitted

$$(A \wedge B) \wedge C = A \wedge (B \wedge C) = A \wedge B \wedge C$$

$$(A \vee B) \vee C = A \vee (B \vee C) = A \vee B \vee C$$

2. Conjunction and disjunction operations have the property of being "commutative"

$$A \wedge B = B \wedge A; \qquad A \vee B = B \vee A$$

3. The so-called "laws of idempotency" are valid for conjunction and disjunction

$$A \wedge A = A; \qquad A \vee A = A$$

4. Conjunction and disjunction operations have the property of being "distributive"

$$(A \vee B) \wedge C = (A \wedge C) \vee (B \wedge C)$$

$$(A \wedge B) \vee C = (A \vee C) \wedge (B \vee C)$$

5. The following relationships are valid

$$A \wedge \overline{A} = 0; \qquad A \vee \overline{A} = 1; \qquad \overline{\overline{A}} = A$$

$$A \wedge 0 = 0; \qquad A \wedge 1 = 1; \qquad A \vee 0 = A; \qquad A \vee 1 = 1$$

6. If we assume that $A^0 = \overline{A}$, $A^1 = A$ and σ_j for any $j = 1, \ldots, n$ is equal to 0 or 1, then the random conjunction

$$A_1^{\sigma_1} \wedge \ldots \wedge A_n^{\sigma_n} = 1$$

when and only when $A_i = \sigma_i$, $i = 1, \ldots, n$. Indeed, the above conjunction will be equal to **1** when and only when all its conjunctive terms equal **1**. Obviously, $A_i^{\sigma_i} = 1$ in that (and only that) case if $A_i = 1$ for $\sigma_i = 1 (A_i^1 = A_i = 1)$ and $A_i = 0$ for $(A_i^0 = \overline{A}_i = 1)$.

The above properties 1–6 remain valid when arbitrary propositional forms are substituted for propositional symbols.

7. If $F(A, B, \ldots)$ is a form containing only logical operations of negation, conjunction, and disjunction, then to obtain the negation $\overline{F}(A, B, \ldots)$, it is necessary to replace all arguments in the input form $F(A, B, \ldots)$ by their negations, while conjunction and disjunction operations should be reversed, i.e., the symbol $\wedge$ should be replaced by $\vee$ and, vice versa, symbol $\vee$ by $\wedge$. This property is known as the theorem of de Morgan.

For instance, if

$$F = (A_1 \vee \overline{A}_2) \wedge (A_3 \vee \overline{A}_4 \vee A_5) \wedge \overline{A}_6$$

then, according to the de Morgan theorem

$$\overline{F} = (\overline{A}_1 \wedge A_2) \vee (\overline{A}_3 \wedge A_4 \wedge \overline{A}_5) \vee A_6$$

From the de Morgan theorem the following implications can be obtained

$$\overline{A \wedge B} = \overline{A} \vee \overline{B}; \qquad \overline{A \vee B} = \overline{A} \wedge \overline{B}$$

The last two properties cannot be verified by means of truth tables.

Normal forms. Using the above properties of logical operations, it is possible to establish a number of theorems which will be used in transforming truth functions. Below, in writing conjunctions, we will be using the sign of a conventional product along with the $\wedge$ sign, i.e., instead of writing $(A_1 \wedge A_2 \wedge \ldots \wedge A_m)$, we will frequently be writing $(A_1 \ldots A_m)$.

Theorem 2.1. Any propositional form $F(A_1, A_2, \ldots, A_k, A_{k+1}, \ldots, A_m)$, except a form that is a contradiction (i.e., identically equal to zero), may be represented as

$$F(A_1, \ldots, A_k, A_{k+1}, \ldots, A_m)$$

$$= \bigvee_{\sigma_1, \ldots, \sigma_k} A_1^{\sigma_1}, \ldots, A_k^{\sigma_k} F_k(\sigma_1, \ldots, \sigma_k, A_{k+1}, \ldots, A_m) \tag{2.1}$$

where the symbol $\bigvee\limits_{\sigma_1, \ldots, \sigma_k}$ denotes logical addition with respect to all sets; $\sigma_1, \ldots, \sigma_k$; $\sigma_i (i = 1, \ldots, m)$ in each set assumes the value 0 or 1.

One can easily verify the validity of this theorem by letting $A_i = \sigma_i (i = 1, \ldots, m)$. Indeed, in this case in the left side of Eq. 2.1, we obtain $F(\sigma_1, \ldots, \sigma_k, A_{k+1}, \ldots, A_m)$.

The representation of a truth function in the format 2.1 is known as an expansion of a function by k variables.

Theorem 2.2. Each propositional form $F(A_1, \ldots, A_m)$, with the exception of the form that is a contradiction, may be represented as

$$F(A_1, \ldots, A_m) = \bigvee_{F(\sigma_1, \ldots, \sigma_m)=1} A_1^{\sigma_1}, \ldots, A_m^{\sigma_m} \qquad (2.2)$$

In the right side of 2.2, the logical addition is performed for all sets $\sigma_1, \ldots, \sigma_m$, for which $F(\sigma_1, \ldots, \sigma_m) = 1$. This theorem directly follows from theorem 2.1. According to theorem 2.2, after expanding the function $F(A_1, \ldots, A_m)$ by m variables, we obtain

$$F(A_1, \ldots, A_m) = \bigvee_{\sigma_1, \ldots, \sigma_m} A_1^{\sigma_1}, \ldots, A_m^{\sigma_m} F(\sigma_1 \ldots, \sigma_m)$$

Since $F(\sigma_1, \ldots, \sigma_m)$ is equal to 0 or 1, then, using property 5, we obtain 2.2 from the last relationship.

The right side of 2.2 is termed the "complete disjunctive normal form" (CDNF) of function $F(A_1, \ldots, A_m)$. Terms of the format $(A_1^{\sigma_1}, \ldots, A_m^{\sigma_m})$ contained in the CDNF are referred to as the "disjunctive terms" of the given CDNF. Thus, a CDNF is a logical sum of a number of disjunctive terms.

From Eq. 2.2, a method for constructing a CDNF of a function given in a tabular form follows. From the table one selects sets $(\sigma_1, \ldots, \sigma_m)$, i.e., sequences consisting of ones and zeros, for which $F(A_1, \ldots, A_m) = 1$. Then, for each set, one assembles a disjunctive term $(A_1^{\sigma_1}, \ldots, A_m^{\sigma_m})$ and then all such terms are connected with disjunction signs.

Suppose, for instance, that a truth function is given in Table 2.2. Preparation of a CDNF of this function is required.

In accordance with the indicated method, we select sequences $(0,0,1)$, $(0,1,1)$, $(1,1,0)$ located in lines 2, 4, and 7, respectively. Using them, we construct the CDNF of the given function as

$$F(A_1, A_2, A_3) = (A_1^0 A_2^0 A_3^1) \vee (A_1^0 A_2^1 A_3^1) \vee (A_1^1 A_2^1 A_3^0)$$

$$= (\overline{A}_1 \overline{A}_2 A_3) \vee (\overline{A}_1 A_2 A_3) \vee (A_1 A_2 \overline{A}_3)$$

Sometimes it is convenient for a truth function to be presented in the format of a logical product of some number of terms $(A_1^{\sigma_1} \vee \ldots \vee A_2^{\sigma_2} \vee \ldots \vee A_m^{\sigma_m})$. Such representation is called the "complete conjunctive normal form" (CCNF). Terms of the format $(A_1^{\sigma_1} \vee A_2^{\sigma_2} \vee \ldots \vee A_m^{\sigma_m})$ contained in a CCNF are referred to as "conjunctive terms."

Theorem 2.3. Any truth form, except the form that is a tautology, may be represented as a complete conjunctive normal form.

Table 2.2

Line Number	A_1	A_2	A_3	F
1	0	0	0	0
2	0	0	1	1
3	0	1	0	0
4	0	1	1	1
5	1	0	0	0
6	1	0	1	0
7	1	1	0	1
8	1	1	1	0

The validity of this theorem directly follows from the method of transition from one normal form of representation of a truth function to another form which will be discussed later.

Rules for forming tautologies and contradictions. Let us consider some rules for forming tautologies and contradictions from disjunctive and conjunctive terms.

1. A logical product of two different disjunctive terms containing n arguments is a contradiction (i.e., identically equal to zero) when and only when one of the terms includes a negation of at least one argument making up the other disjunctive term. For instance

$$(A_1 A_2 A_3)(\overline{A}_1 A_4 A_5) = 0$$

2. A logical sum of two different conjunctive terms containing m arguments is a tautology when and only when one of the terms contains a negation in at least one of the arguments contained in the other term. For instance

$$(A_1 \lor A_2 \lor \overline{A}_3) \lor (A_4 \lor \overline{A}_5 \lor A_3) = 1$$

These rules follow directly from the property of logic operations #5. Obviously, these rules remain in effect if the conjunctive terms in the first case, and disjunctive terms in the second case, contain a different number of arguments.

3. A logical sum of all 2^m disjunctive terms of m arguments, each pair of which is different, is a tautology

$$\underset{\sigma_1, \ldots, \sigma_m}{\lor} A_1^{\sigma_1} \ldots A_m^{\sigma_m} = 1 \tag{2.3}$$

The validity of this rule may be proven by means of mathematical induction. Indeed, for $m = 1$, Eq. 2.2 becomes

$$A_1 \vee \overline{A}_1 = 1$$

which follows from the property of logical operations #5. Let Eq. 2.3 be valid for $m = k$

$$\bigvee_{\sigma_1,\dots,\sigma_k} A_1^{\sigma_1} A_2^{\sigma_2} \dots A_k^{\sigma_k} = 1$$

For $m = k + 1$, then we have

$$\bigvee_{\sigma_1,\dots,\sigma_{k+1}} A_1^{\sigma_1} \dots A_{k+1}^{\sigma_{k+1}}$$

$$= \left(\bigvee_{\sigma_1,\dots\sigma_k} A_1^{\sigma_1} \dots A_k^{\sigma_k} \right) A_{k+1} \vee \left(\bigvee_{\sigma_1,\dots,\sigma_k} A_1^{\sigma_1} \dots A_k^{\sigma_k} \right) \overline{A}_{k+1}$$

$$= A_{k+1} + \overline{A}_{k+1} = 1$$

4. A logical product of 2^m conjunctive terms of m arguments in each pair different is a contradiction

$$\bigwedge_{\sigma_1,\dots,\sigma_m} (A_1^{\sigma_1} \vee \dots \vee A_m^{\sigma_m}) = 0 \qquad (2.4)$$

Here the sign $\bigwedge_{\sigma_1,\dots,\sigma_m}$ denotes the logical multiplication of all sets $(\sigma_1,\dots,\sigma_m)$. To prove this rule, let us introduce notations for disjunctive and conjunctive terms

$$\left. \begin{array}{l} D_i = A_1^{\sigma_1} \dots A_m^{\sigma_m} \\ K_j = A_1^{\overline{\sigma}_1} \vee \dots \vee A_m^{\overline{\sigma}_m} \end{array} \right\} \qquad (2.5)$$

where subscripts i and j are respectively equal to binary numbers $\sigma_1 \dots \sigma_m$ and $\overline{\sigma}_1,\dots,\overline{\sigma}_m$

$$i = \sigma_1 \dots \sigma_m = \sigma_1 2^{m-1} + \sigma_2 2^{m-2} + \dots + \sigma_m 2^0$$

$$j = \overline{\sigma}_1 \dots \overline{\sigma}_m = \overline{\sigma}_1 2^{m-1} + \overline{\sigma}_2 2^{m-2} + \dots + \overline{\sigma}_m 2^0$$

Let number l be dual to the number k if, when recording them in binary form, the number l may be obtained from k by replacing in this latter 1's by 0's and 0's by 1's. Obviously, if the number l is a dual to the number k, then k is dual to l. Let in 2.5 $\sigma_i = 1$, if $\overline{\sigma}_i = 0$, and $\sigma_1 = 0$, if $\overline{\sigma}_i = 1(i = 1,\dots,m)$. Then, apparently, from the de Morgan theorem

$$\overline{D}_i = K_j; \qquad \overline{K}_j = D_i \qquad (2.6)$$

i.e., the negation of the disjunctive term D_i is equal to the conjunctive term K_j; and negation of the conjunctive term K_j is equal to the disjunctive term D_i, and subscripts i and j are mutually dual numbers.

Let us prove rule 4. To do this, Eq 2.3 is subjected to the logical operation of negation. Then, according to the de Morgan theorem, we get

$$\underset{\overline{\sigma}_1,\dots,\overline{\sigma}_m}{\wedge} (A_1^{\overline{\sigma}_1} \vee \dots \vee A_m^{\overline{\sigma}_m}) = 0$$

Since logical multiplication is performed by all sets $(\overline{\sigma}_1, \dots, \overline{\sigma}_m)$, then the last relationship coincides with Eq. 2.4, which is exactly what was required.

Transition from one normal form of function representation to another. It was shown above that any propositional form may be represented in CDNF. A method for constructing a CDNF for a function given in tabular form was described. Therefore, let us consider a method for changing the representations of a function from CDNF to CCNF.

Let us write Eq. 2.3, using notation of Eq. 2.5

$$\overset{2^m-1}{\underset{i=0}{\vee}} D_i = 1 \tag{2.7}$$

The symbol $\overset{2^m-1}{\underset{i=0}{\vee}}$ denotes logical summation from 0 to $2^m - 1$. Suppose we are given a function $F(A_1, \dots, A_m)$ in CDNF format

$$F(A_1, \dots, A_m) = D_{i_0} \vee D_{i_l} \vee \dots \vee D_{i_k} \tag{2.8}$$

where $D_{i_l} (l = 0, 1, \dots, k)$ is the disjunctive terms.

Since $\overline{F} \vee F = 1$, then from Eq. 2.7 it follows that

$$\overline{F} = \overline{F}(A_1, \dots, A_m) = \underset{i \neq i_0, i_1, \dots, i_k}{\wedge} D_i$$

In this relationship, the logic summation is carried out for all i in the range between 0 and $2^m - 1$, which are not equal to $i_0, i_1, \dots, i_k$. In other words, disjunctive terms contained in 2.8 are excluded. Using the de Morgan theorem, from the last relationship we obtain a representation of function F in CCNF

$$F(A_1, \dots, A_m) = \overline{\overline{F}} = \underset{j \neq \overline{i}_0, \overline{i}_1, \dots, \overline{i}_k}{\wedge} K_j$$

where $\overline{i}_l$ denotes a number dual to $i_l (l = 0, 1, \dots, k)$; $\underset{j \neq \overline{i}_0, \overline{i}_1, \dots, \overline{i}_k}{\wedge}$ is the logical multiplication by all that are not equal to $\overline{i}_l (l = 0, 1, \dots, k)$ in the range between 0 and $2^m - 1$.

Thus, to achieve a transition from CDNF to CCNF, one should:

1) Form a logical sum of the disjunctive terms not contained in CDNF;
2) Replace arguments by their negations, the conjunction sign — by the disjunction sign, and the disjunction sign — by a conjunction sign.

Suppose, for instance, that we are given a function $F(A_1, A_2, A_3)$ represented in CDNF format

$$F(A_1, A_2, A_3) = \overline{A}_1 A_2 \overline{A}_3 \vee A_1 \overline{A}_2 A_3 \vee A_1 A_2 \overline{A}_3$$

Since $m = 3$, there will be eight (2^3) various disjunctive terms $D_0, D_1, \ldots, D_7$. The CDNF representation contains disjunctive terms D_{i_0}, D_{i_1} and D_{i_2}, where $i_0 = 010 = 2$, $i_1 = 101 = 5$, and $i_2 = 110 = 6$, i.e.

$$F(A_1, A_2, A_3) = D_2 \vee D_5 \vee D_6$$

Consequently

$$\overline{F}(A_1, A_2, A_3) = D_0 \vee D_1 \vee D_3 \vee D_4 \vee D_7$$

$$= \overline{A}_1 \overline{A}_2 \overline{A}_3 \vee \overline{A}_1 \overline{A}_2 A_3 \vee \overline{A}_1 A_2 A_3 \vee A_1 \overline{A}_2 \overline{A}_3 \vee A_1 A_2 A_3$$

Applying a negation to the last equation, i.e., replacing arguments by their negations, conjunctive signs by disjunctive signs and vice versa, we obtain

$$F(A_1, A_2, A_3) = (A_1 \vee A_2 \vee A_3)(A_1 \vee A_2 \vee \overline{A}_3)$$

$$\times (A_1 \vee \overline{A}_2 \vee \overline{A}_3)(\overline{A}_1 \vee A_2 \vee A_3)(\overline{A}_1 \vee \overline{A}_2 \vee \overline{A}_3)$$

which determines the representation of the input function in the CCNF format.

Synthesis of finite automata. The tools of the propositional calculus (logic algebra) are widely used in the synthesis and analysis of finite automata. Here the propositional characters denote the state of switching elements. It is commonly assumed that a propositional character equal 1 if its corresponding contact is closed, and 0 if it is open. The above theory is used in the synthesis of switching systems.

Suppose that in an n-story building we wish to install switches in such a manner, that by switching any one of them, stairway lights on every floor are turned ON if they were OFF previously, or are turned OFF if they were ON previously.

Let us take a case with $n = 2$. The state of a switch at the i-th floor will be denoted by $B_i (i = 1, 2)$. Let us assume that $B_i = 1$ if the switch at the i-th floor is ON, and $B_i = 0$ if it is OFF. The state of the light will be denoted by C, with $C = 1$ when a light is on and $C = 0$ when a light is OFF. The two switches may form four possible states (Table 2.3).

Suppose that in state 1 the light is OFF ($C = 0$). Then, if one of the switches is turned ON (state 2 or 3), the light should go ON ($C = 1$). If the other switch is turned ON (state 4), then the light should go OFF ($C = 0$). Thus, for $C = C(B1, B2)$, we obtain the truth table 2.3.

Using the rule for constructing a CDNF function on the basis of tabular data, we get

$$C = \overline{B}1B2 \vee B1\overline{B}2$$

Table 2.3

State of Switches	B1	B2	C
1	0	0	0
2	0	1	1
3	1	0	1
4	1	1	0

The circuit of this function is depicted in Fig. 2.1.

More involved and meaningful examples of analysis and synthesis of finite control systems are treated in books on theory of finite automata [2].

The calculus of predicates. A predicate or a logical function is a function with any number of arguments that assume truth values T and F (**1** and **0**). The arguments receive values from a random finite or infinite set **M** designated as the "object domain." A predicate with n arguments is called the n-place predicate. If $F(x)$, $G(x, y)$, $\mathcal{P}(x_1, \ldots, x_n)$ are predicates, then characters F, G and $\mathcal{P}$ denoting the function are called "predicate characters," and their arguments x, y, x_1, $\ldots$, x_n "object variables." A common expression is that "a predicate $F(x)$ is obtained by applying a predicate character F to an object variable x."

Instead of object variables, one can substitute into predicates specific values from an object domain **M** — object constants, as well as **n**-place functions $f(x_1, \ldots, x_n)$ reflecting $\mathbf{M}^n$ in **M**, i.e, functions that receive values from **M** and defined in $\mathbf{M}^n$. Such functions are termed functions in **M**. In other words, predicate characters can also be applied to object constants and functions in **M**. Usually it is said about function $f(x_1, \ldots, x_n)$ that it was obtained as a result of the application of an n-place functional character to predicate variables $x_1, \ldots, x_n$.

Predicate $F(x)$ defined in an object domain **M** assigns a specific property to the elements of a set **M** and is interpreted as a statement "x has the property F." Here $F(x)$ assumes the value T if this proposition is true or F is this proposition is false. The predicate $F(x_1, \ldots, x_n)$ determines the relation among the elements $x_1, \ldots, x_n$ and is interpreted as notation of the statment "$x_1, \ldots, x_n$ are related to each other by F." Suppose, for instance, that **M** is a set of positive integers. Then the predicate $F(x)$ may mean that "x is an even number" or "x is an odd number," or "x is a prime number;" a predicate $G(x, y)$ may mean "x is larger than y" or "x is not larger than y," etc.

Let us denote predicate characters by capital characters F, G, H, P, and Q; functional characters — by lower case letters f, g, h, p, q; object variables — by lower case characters from the end of the alphabet x, y, z, and w; object constants

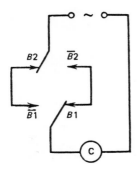

Figure 2.1 Function realization circuit.

— by lower case characters from the beginning of the alphabet a, b, c, ... All the above characters will also be used with subscripts.

Quantors and bound variables. Truth functions are a special case of predicates when their arguments assume only two values, **0** and **1** (T and F). All operations of the propositional calculus are transferred to the predicate calculus and are used to relate predicates and formulae (formulae will be defined later). But in the predicate logic, for a more compact recording of propositions like "for each x, $F(x)$ is true" and "there exists such an x for which $F(x)$ is true," two additional operations are introduced, "quantor of commonality $\forall$" and "quantor of existance $\exists$." With the aid of these operations, the above propositions are written as $\forall x\, F(x)$ and $\exists x\, F(x)$, respectively. The expression $\forall x\, F(x)$ denotes a true proposition when $F(x)$ is true for all $x \in M$ and false when it is otherwise. The expression $\exists x\, F(x)$ denotes a true proposition when there exists an element $x \in M$ for which $F(x)$ is true, and it is false otherwise. If $F(x)$ in actuality does not depend on x, then the expressions $\forall x\, F(x)$ and $\exists x\, F(x)$ have the same meaning as $F(x)$. Quantors $\forall$ and $\exists$ are called dual.

Let us consider expression $\overline{\forall x\, F(x)}$, i.e., a negation of expression $\forall x\, F(x)$. This expression denotes the proposition "$\forall x\, F(x)$ is false," that is equivalent to the proposition "there exists an element x for which $F(x)$ is false" or, which is the same, "there exists an element x for which $\overline{F}(x)$ is true." Consequently, the expression $\overline{\forall x\, F(x)}$ is equivalent to the expression $\exists x\, \overline{F}(x)$

$$\overline{\forall x\, F(x)} = \exists x\, \overline{F}(x) \qquad (2.9)$$

The expression $\overline{\exists x\, F(x)}$ which is a negation of $\exists x\, F(x)$ and which says "$\exists x\, F(x)$ is false," is equivalent to a proposition "for all x the proposition $F(x)$ is false" or, which is the same, "for all x the proposition $F(x)$ is true." Consequently, the expression $\overline{\exists x\, F(x)}$ is equivalent to the expression

$$\overline{\exists x\, F(x)} = \forall x\, \overline{F}(x) \qquad (2.10)$$

It is commonly accepted that quantors $\forall$ and $\exists$ in expressions $\forall x F(x,y)$, $\exists x F(x,y)$ apply to the variable x or that the variable is linked to a corresponding quantor. A variable that is linked to a quantor is designated as "bound," while a variable not linked to any quantor is called "unbound," or "free."

The main object of investigations into the logic of predicates is formula. In its definition, a concept "therm" is used that combines the name of elements and functions to which the predicate character applies.

A "therm" is recursively defined as follows:

- Every object variable or object constant is a therm;

- If f is an n-place functional character and $t_1, \ldots, t_n$ are therms, then $f(t_1, \ldots, t_n)$ is a therm;

- No other expressions are therms.

Let us introduce a concept of an "elementary formula:"

- Every propositional character is an elementary formula;

- If F is an n-place predicate, and $t_1, \ldots, t_n$ are therms, then $F(t_1, \ldots, t_n)$ is an elementary formula;

- No other expressions are elementary formulae.

Elementary formulae are sometimes called "atomic formulae." "Formulae" in the predicate calculus are defined as follows:

- Every elementary formula is a formula.

If $\mathcal{F}$ and $\mathcal{G}$ are formulae, and x is an object variable contained in $\mathcal{F}$, then each of the expressions $(\forall x \mathcal{F})$, $(\exists x \mathcal{F})$, $\overline{\mathcal{F}}$, $(\mathcal{F} \wedge \mathcal{G})$, $(\mathcal{F} \vee \mathcal{G})$, $(\mathcal{F} \rightarrow \mathcal{G})$ is a formula;

- No other expression is a formula.

Sometimes instead of formula, the expression "tamely constructed formula" (TC-formula) is used. Examples of TC-formulae are the following expressions

$$F(a, x, g(y, a, b)), \ F(x, y) \rightarrow G(a, b)$$

$$(F(a) \rightarrow (F(a) \vee \overline{G}(b)))$$

Expressions $\overline{f}(x), f(F(b)), F(f(a), (G(b) \rightarrow H(c)))$ are not formulae.

Interpretations. A formula has a specific meaning, i.e., it stands for a specific proposition, if there exists some interpretation. To interpret a formula means to relate to it some nonempty set $\mathbf{M}$, i.e., to specify an object domain (also called an "interpretation domain") and to specify a correlation that relates:

- to each object constant in the formula a specific element from **M**;

- to each n-place functional character in the formula a specific n-place function in **M**;

- to each n-place predicate character in the formula a specific relationship among n elements from **M**.

In other words, an interpretation of formulae of the predicate calculus is a concrete definition of an object domain **M** and correlation between the symbols (object constants, functional and predicate characters) entering the formula on the one hand, and elements, functions, and relationships or **M** on the other.

Let us consider an elementary formula

$$G(f(a,b),\ g(a,b))$$

and the following interpretation:

- **M** is a set of real numbers;

- $a,\ b$ are numbers 2 and 3, respectively;

- f is a summing function $(f(a,b) = a + b)$;

- g is a multiplication function $(g(a,b) = ab)$;

- G is a relationship "not less."

In this interpretation, the above formula stands for the proposition "the sum $2 + 3$ is not less than the product $2 \cdot 3$." This proposition is incorrect and therefore $G(f(a,b),\ g(a,b)) = F$. If this interpretation is modified by assuming $b = 1$ or $b = 2$, then $G(f(a,b),\ g(a,b)) = T$. Obviously, there are many other interpretations for which the above formula in some cases has the value F and in others, the value T. But there is no interpretation for which it simultaneously has the values T and F.

Let us consider another elementary formula

$$G(f(g(x,x),\ g(y,y)),\ g(a,g(x,y)))$$

with the same interpretation. This formula represents the proposition "$x^2 + y^2 \geq 2xy$," which is correct for any x and y from **M** and has the value of T.

For a given interpretation, i.e., for describing propositions using the calculus of predicates language in a given object domain, for denoting predicate characters and object variables or constants, words (or their abbreviations) that are names of properties, correlations, and objects which they define are frequently used. For instance, propositions "a book is on the table" and "Smith is either at home or at work," may be written in the calculus of predicate language as AT (book, table) and AT (smith, home) $\lor$ AT (smith, work).

General validity, equivalence, and infeasibility. A formula without free variables is called a "closed formula." For a given interpretation, every closed formula is a proposition that is true or false, while every formula with free variables expresses some relationship over the interpretation domain that may be true for certain values of variables from the interpretation region and false for others.

If some formula has a value T for all interpretations, then it is referred to as one with "general validity." Examples of general-validity formulae are $(\mathcal{F} \vee \overline{\mathcal{F}})$, $(\mathcal{F} \rightarrow (\mathcal{F} \vee \mathcal{G}))$.

Two formulae $\mathcal{F}$ and $\mathcal{G}$ are "equivalent" if the formula $(\mathcal{F} \equiv \mathcal{G})$ is generally valid, i.e., for all interpretations the value $\mathcal{F}$ is true when and only when the value of $\mathcal{G}$ is true. Examples of equivalent formulae are $\mathcal{F}$ and $\overline{\overline{\mathcal{F}}}$, $(\overline{\mathcal{F} \vee \mathcal{G}})$ and $(\overline{\mathcal{F}} \wedge \mathcal{G})$. If $\mathcal{F}$ and $\mathcal{G}$ are equivalent formulae, then we will write $\mathcal{F} = \mathcal{G}$.

If for a given interpretation a formula $\mathcal{F}$ has the value T, this means that the interpretation satisfies formula $\mathcal{F}$. If some set of formulae is given and if for a given interpretation each formula within this set has the value T, then this interpretation satisfies this set. The formula logically follows from some set of formulae Φ_0 (denoted $\Phi_0 \Rightarrow \mathcal{F}$) if every interpretation that satisfies Φ_0 also satisfies $\mathcal{F}$.

If some set of formulae is not satisfied by any interpretation, then such a set is termed "unsatisfactory" or "infeasible." If, for instance, $\mathcal{F}$ logically follows from Φ_0, then the combination $\Phi_0 \cup \overline{\mathcal{F}}$, i.e., a set comprised of formulae of set Φ_0 and formula $\overline{\mathcal{F}}$, is infeasible.

If a set Φ is not infeasible, i.e., there exists some interpretation that satisfies Φ, then the set Φ is called called "feasible."

Additional information on calculus of propositions and predicates and their applications in different areas may be found in [3, 4].

2.3   SEMANTIC NETWORKS

The basis of the formalization of semantic knowledge about an object domain is often a directed graph with marked vertices and arcs, known as a semantic network. The vertices correspond to specific objects, while arcs correspond to correlations between them. Markings of vertices have a reference character and represent some names. Such names, for instance, may be words in the natural language. Markings of arcs denote elements of a correlation set.

Let us consider a more rigorous definition of a semantic network [1]. Suppose we are given finite sets of symbols $A = \{A_1, \ldots, A_4\}$ called attributes, and a finite set of correlations $R = \{R_1, \ldots, R_m\}$. A configuration or an intensional of correlation R_i is a set of pairs

$$\mathrm{INT}(R_i) = \{\ldots, [A_j, \mathrm{DOM}(A_j)], \ldots\} \tag{2.11}$$

where R_i is the name of the correlation; $\mathrm{DOM}(A_j)$ is the domain A_j, i.e., the set of values of the attribute A_j of the correlation R_i.

The combination of all domains is called "base set of a model" or a "set of objects" on which the relationships R_i are given.

The extensional of the relationship R_i is a set

$$\mathrm{EXT}(R_i) = \{F_1, \ldots, F_p\}$$

where $F_k (k = \overline{1, p})$ is the fact of the correlation R_i. A fact is assigned by a multiplicity of pairs of the form "attribute-value" referred to as attribute pairs. The term "fact" is a process of defining a specific relationship between the objects in question. In a graphic interpretation, a fact is a star-shaped subgraph of a semantic network. The root of a subgraph is a vertex of a predicate type marked by a universal marking containing the name of a corresponding correlation. From a vertex of a fact, there originate edges marked by attribute names of the given fact and directed toward the vertices of the base set which are the values of these attributes.

Let us consider an example demonstrating the introduced definitions. Suppose we are given a base set of a model (a set of objects) which is a combination of integers $\{0, 1, 2\}$; a set of correlation R_i that contains two correlations "larger-smaller" and "sum" (denoted "<" and "+," respectively). The intensional of these relationships will be, obviously, written as follows:

INT $(+) = \{$ [1st addend, (0, 1, 2), [2nd addend, (0, 1, 2)], [sum, (0, 1, 2)]$\}$;
INT $(<) = \{$ [smaller, (0, 1)], [larger, (1, 2)]$\}$.

The extensionals of these relationships may be written in the format of facts, for instance, as follows:

$F$1: (<, smaller 0, larger 0 1);
$F$2: (<, smaller 1, larger 2);
....
$F$9: (+, 1st addend 2, 2nd addend 0, sum 2).

Sometimes it is convenient to use a table (Tables 2.4, 2.5).

A graphic representation of an extensional semantic network is shown in Fig. 2.2. The following rules and notation were used in its construction. The base set of the model corresponds to vertices 0, 1, 2, denoted by squares. Facts vertices are denoted by circles with markings formed by pairs of identifiers: a fact and a correlation. Arcs carry markings of corresponding attributes.

The extensional semantic network describes extensional knowledge about modeled objects; it is effectively a "photograph" of its current state. An intensional semantic network contains intensional knowledge and describes the overall structure of the modeled object region on the basis of abstract objects and relationships, i.e., generalized representatives of some classes of objects and relationships. For instance, such objects as a MANUFACTURING SECTOR, LOAD, PART may be generalized concepts whose value set forms a set of names of specific manufacturing sectors (turning shop, stamping shop), a set of names of loads (cassette, blank), a set of names of parts (bolt, shaft, chassis).

Table 2.4

Fact marker	Corrrelation "less"	
	1	2
$F1$	0	1
$F2$	1	2
$F3$	0	2

There are three main types of objects that are used in semantic networks: concepts, events, and properties.

"Concepts" represent information about abstract or physical objects of an object domain. General concepts are interpreted as sets of domains (of parameters or constants).

"Events" are operations that may affect changes in an object domain. A result of an event is some new state of the object domain. One may designate some desired (target) state of an object domain and pose a problem of determining in the semantic network of a sequence of events that would bring about the target state.

"Properties" are used for the refinement of concepts, events or other properties. As applied to concepts, properties describe their peculiarities or characteristics: color, dimensions, quality; as applied to events, properties describe duration, location, time, etc.

The variety of semantic correlations is conditionally divided into four classes: linguistic, logical, set-theoretic, and quantified [5].

Table 2.5

Fact marker	Correlation "sum"		
	1st addend	2nd addend	sum
$F4$	0	0	0
$F5$	0	1	1
$F6$	0	2	2
$F7$	1	0	1
$F8$	1	1	2
$F9$	2	0	2

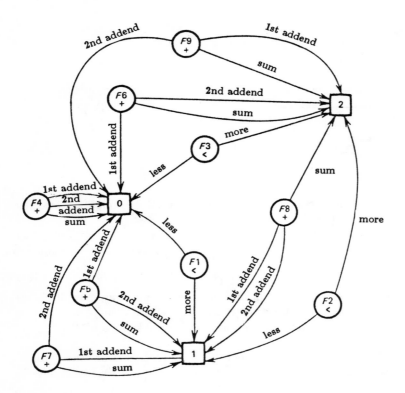

Figure 2.2   Extensional semantic network.

Linguistic correlations. The most widely used linguistic correlation are case correlations that, includes, among other things: an agent — a correlation between an event and its cause; an object — a correlation between an event and the object subjected to an action; a condition – a correlation defining the logical interdependence between events; a tool — an object with whose help the event occurs; a place — the place where the events are occurring. Another type of linguistic correlations is the characterization of verbs and attribute relationships. Verbs are characterized by their mood, gender, voice, plural or singular, time. Attribute relationships include color, dimensions, shape, modifications, etc. For instance, a sentence "Large red balloons" with the aid of attribute relationships may be represented by a diagram of Fig. 2.3.

Logical correlations are operations that are used in the propositional calculus: disjunction, conjunction, implication, negation.

Set-theory correlations are a subset (denoted by SUB), superset (denoted by SUP), an element of a set, a relationship between a part and the whole, etc. This class of correlations is used for constructing hierarchial taxonomic (coordinative) structures for representing generalized information. Figure 2.4 depicts a fragment of a taxonomic structure of the concept "robots." In such structures all properties of a SUBconcept automatically are assigned to a SUPconcept. This permits one to exclude repetitive information from the database.

Quantified correlations are logical quantors of generality and existence. Logical quantors are used to represent knowledge of declarative type for recording, for instance, such statements as "Every machine tool needs preventive maintenance" or "There is a robot A that can service all machine tools within group B." A formal record of such statements is based on the predicate calculus language of the first order.

As an example of a semantic network, let us consider a representation of knowledge contained in the sentence "If a machine tool completed a job, the robot loads the cassette with parts into a robocar (AGV), which carries them to the warehouse where a stacker places the cassette into a bin." Let us isolate five factors: the machine tool completed a job ($F1$), the robot loads ($F2$), the robocar transports ($F3$), the cassette contains ($F4$), the stacker distributes ($F5$). Let us note that in describing phrases in natural language, facts are often referred to as propositions. As in the previous example, facts will be denoted by circles and their associated concepts by rectangles. Arcs will be marked by the names of the relationships they express. The diagram of the semantic networks is shown in Fig. 2.5.

An important concept in semantic networks is the "designate." This is a unique intrasystem designation that corresponds to some object of the object domain if information about it is lacking at a given time. A designate reflects the most significant meaning of the object, such as the fact of its existence. For instance, a sentence of the type "The machine tool $C1$ has a buffer" contains an uncertainty about the buffer characteristics. Let us assume that as information is accumulated (this may occur during the design process), its capacity, dimensions, etc. will become defined. Then the above sentence might be represented as follows:

($F1$: HAS agent MACHINE TOOL object $D1$)
($D1$: name BUFFER)
($D1$: size _____)

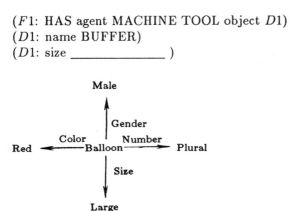

Figure 2.3 Diagram illustrating attributive relationships.

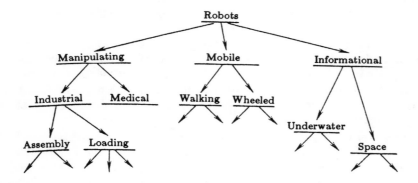

Figure 2.4 Fragment of a taxonomial structure of the concept "robots."

($D1$: capacity _____)

Here $F1$ is the fact marker; $D1$ is the designate marker. Blanks will be filled by values of the associated attribute as the data is accumulated. Thus, at the first mention of the object, its designate is entered into the database that may be manipulated without waiting for all information about the object. The introduction of designates increases the flexibility of a semantic network. Their utilization, however, requires additional computing resources.

Let us consider certain aspects in developing intelligent data banks on the basis of semantic networks. The main principle of their organization is the separation of extensional and intensional knowledge, keeping in mind that the extensional

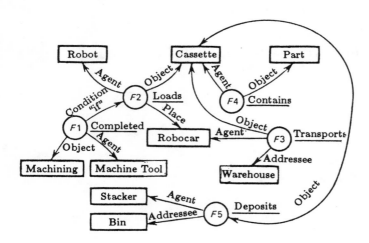

Figure 2.5 A semantic network.

semantic network is the foundation of the data base, while the intensional semantic network is the foundation of the knowledge base. The "delamination" of the system of knowledge representation substantially increases its expressiveness potential in, for instance, logical models. The diagram of a system of knowledge representation (SKR) of this type is depicted in Fig. 2.6. The conditionality of this diagram is due to the absence of a strictly-defined separation of the above-mentioned (SKR) components during their realization.

Information retrieval is the major operating mode of SKR. A request consists of a set of facts (a situation) in the description of which one is allowed to use variables in positions of attribute values, attributes, and names of relationships. A request may be represented by a graph whose vertices corresponding to certain variables are not defined. The search for an answer consists of solving the problem of how to isomorphically include the request graph (or its subgraph) into a semantic network. There are two request types: those concerned with existence, and those concerned with enumeration. The existence request does not contain variables and expects the answer YES if the isomorphic inclusion of the request graph was successful, and expects a NO in the opposite case. During the processing of the enumeration request, a search (enumeration) (for the request graph in question) of all possible isomorphic subgraphs in the semantic network, as well as for defining the variables is performed.

Suppose, for instance, there is the request "The operator reported that the robocar is transporting something. Determine what the robocar is carrying and to what destination." The search for the answer is performed in a semantic network shown in Fig. 2.7, where vertices X and Y have to be defined and where $F6$ is the fact of the reporting. Obviously, the star-shaped subgraph of the fact F in the request graph (identified by a dashed line in the figure) may be isomorphically included into the semantic network if it is made to coincide with $F3$ of Fig. 2.5. In doing so, vertices X and Y will be defined: X is the CASSETTE, Y is the WAREHOUSE. The answer will be as follows: THE ROBOCAR TRANSPORTS THE CASSETTE TO THE WAREHOUSE. This example demonstrates a search

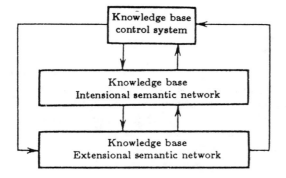

Figure 2.6   Knowledge representation system.

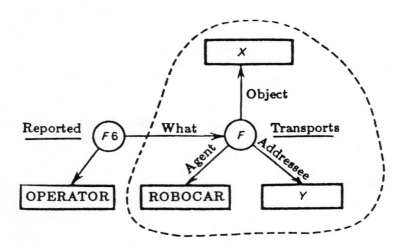

Figure 2.7   Request graph.

for an answer in a semantic network in a simplified form, since here we are not concerned with the process of interpreting the request, initially expressed in a special report language.

A semantic network may be employed for representing both declarative and procedural knowledge. Representation of the declarative knowledge uses the so-called base relationships that are characterized by the fact that their extensional is stored in the database in its entirety. Procedural knowledge elements are linked by virtual relationships that do not have an extensional stored in an explicit form. In the description of these relationships there are rules (programs) that permit the construction of facts as required. Virtual relationships are realized in the form of computable relationships that have a functional or predicate nature either in the form of rules that utilize the properties of symmetry, reflectiveness, or transitiveness (in those cases when they occur in the modeled object domain).

2.4   FRAMES FOR KNOWLEDGE REPRESENTATION

The term "frame" was offered as a designation for a description of some object or phenomenon that has the following property: if any part of it is removed from the description, properties that characterize the object are lost.

A frame is most frequently defined as a data structure for representing stereotype situations. Here the method of classifying a set of specific situations as a stereotype set cannot, as a rule, be strictly defined. In most cases, what is and what is not stereotype is determined by the researcher on the basis of experience and experimental data. Nonformal knowledge of a researcher about the object domain may be viewed as a system of concepts determining his understanding of a specific situation. Every concept is related to a specific situation and these, in turn, are coordinated with a corresponding stereotype situation. While the concepts represent a nonformal knowledge about the stereotype situation, the frames represent formalized knowledge. Thus, frames correspond to concepts that are reflecting objects, phenomena, and characteristics of an object domain. This gives a basis for viewing a frame as a semantic block or module of a model for representing knowledge. A model of knowledge representation is constructed in the form of a network of frames, i.e., a system of frames that are interrelated in a specific manner. Therefore, in knowledge representation models constructed on the basis of frames, one identifies two individual parts: a set of frames that make up a library of internal knowledge representation and a mechanism for their transformation, interlinking, etc.

In general, a frame contains both informational and procedural elements that perform the transformation of information within the frame and its communications with other frames. An important feature of frames is the presence of slots (voids, crevices) in the informational and procedural elements. The slots may be filled in the process of frame activation in accordance with specific conditions. This gives the property of adaptability to a model of knowledge representation both at the module level and at the level of the entire frame network.

Thus, frames are declarative-procedural structures, i.e., arrays of descriptions and (in some cases) of procedures related to them that can be accessed directly from a frame.

A large number of concepts, definitions, and models of frames exists. There are not only formats or records and representation of frames, but also, to some degree, their contents. In a most general case, a frame is defined as a knowledge representation structure of the following format [1]

$$\{n, \ (v_1, g_1, p_1), \ (v_2, g_2, p_2), \ldots, \ (v_k, g_k, p_k)\}$$

where n is the frame name; v_j is the slot name; g_i is the slot value; p_j is the procedure.

A procedure is a permissible, but not necessary, element of a slot. Frame names are used as mnemonic elements for constructing frame networks. Names of other frames may be used as slot values. This ensures linkage among frames as well as their "insertability" into each other.

There are two types of frames: a frame-prototype and a frame-sample. A *frame-prototype* is an intensional description of a set of frame-samples; a *frame-sample* is an extensional description of a corresponding frame-prototype.

Let us consider the concept of frames utilized in a dialogue system GUS. Frame structure here has the following format:

<frame>:: = [<name>]<prototype reference>
<slot> {<slot>}

Square brackets denote the optional character of the element since a name may be absent in a frame-sample; figure brackets denote some set (possibly empty) of elements; the sign ::= is reading as "by definition is." Slot structure is given by:

<slot>:: = <slot name><value> [{<procedure>}]

As an example, let us consider a simplified configuration of a frame-prototype of the concept DATA [3, Chapter 1]:

<DATA> (<MONTH><NAME>)
(<DAY><INTEGERS {1, 2 . . . 31} >)(<YEAR><function>)
(<DAY OF THE WEEK><listing {<MON, TUE, . . . , SUN} ><function>)

The name of the frame-prototype is DATE. In the slot MONTH, the value is designated as NAME, i.e., any character expression may serve as the slot value. Value of the slot DAY may be an integer and their list is given in the slot. Any LISP language function can serve as a function in GUS (see §2.6). Thus, in the slot YEAR, the following procedures may be set up with the use of LISP. If the year is given in the input statement, it is entered into the value field of the slot in the frame-sample; if the year is not given, a procedure is activated that substitutes a current year for the missing value. Such functions are termed "default functions."

In the slot DAY OF THE WEEK, with the aid of LISP, one may set up, for instance, procedures that during the processing of an input statement will be called up automatically for checking the noncontradictory character of the value of the day of the week specified by the user, or for a computation of this value if the user did not specify it.

A specific frame-sample of the frame DATE may be as follows:

<ISA DATE> (<MONTH><JUNE>>) (<DAY>< 5 >)

The marker ISA means that this frame is a frame-sample. Only two slots are filled above. Values of the remaining slots may be computed with the aid of appropriate procedures.

Procedures that are included into a slot are divided into two types: *procedures-demons* and *procedures-slaves*.

Procedures-demons are activated automatically every time whenever data is entered or removed from an appropriate frame-sample. Thus, the procedure built into the DAY OF THE WEEK slot in the above example is a typical example of a procedure-demon. With the aid of procedures of this kind, all routine operations related to the entry of data and knowledge bases are automatically performed. Suppose, for example, that a database describes a transportation-warehouse system in which the same parts are stored at different locations. With the aid of procedures-demons, the communications with the system may be set up in such a way that the user will be specifying only a destination for each lot of parts. A corresponding procedure-demon is activated automatically as soon as the slot ADDRESSEE of the appropriate frame is filled. As a result of performing a

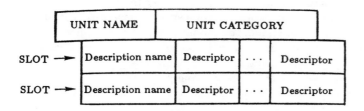

UNIT NAME	UNIT CATEGORY		
Description name	Descriptor	. . .	Descriptor
Description name	Descriptor	. . .	Descriptor

SLOT →
SLOT →

Figure 2.8   Frame structure in the KRL system.

procedure, the name of the given lot will be removed from the frame describing a previous storage location and will be added to the frame of the new storage location.

Procedures-slaves are activated only in response to a request. An example of such a procedure is a function built into the slot YEAR in the frame-prototype DATE. It is called only when the user did not indicate the year.

Experiments with the system GUS stimulated the development of new programming techniques for representing knowledge of the type KRL, FRL, and others based upon frame concepts that differ from those of GUS.

Knowledge description in system KRL is done with the aid of the so-called units that have a frame-like structure (Fig. 2.8). The unit (frame) is characterized by a category. Within the KRL there are such categories as main, abstract, individual, correlation, statement, and a number of other categories. Categories define methods of information processing in the system. For instance, when a pair of frames is compared during an information search, the noncoincidence of their categories permits omission of the analysis of the internal elements of their structures.

As in GUS, slots (descriptions) in KRL may have linked procedures set up by means of LISP functions. A slot in KRL may include several descriptors permitting the characterization of an object from different points of view, such as a class, a category, a relationship in which the object takes part. Altogether, there are 12 types of descriptions in KRL. Their utilization significantly broadens the capabilities of representing object properties.

The concept of frames employed in the FRL system provides wide-ranging capabilities for developing hierarchial frame structures. This is achieved by assigning one slot of the frame for referring to the previous (within the hierarchy) frame. In this way, the "insertability" of frames is ensured. Up to five levels of "insertability" are allowed. These properties permit the development of a mechanism of the so-called property inheritance.

Let us explain the operation of the inheritance mechanism by means of an example. Suppose that storage addresses of a large number of parts coincide almost entirely. They have the same warehouse number, the same section number within the warehouse, and the same block number within the section. Only bin numbers are different. Obviously, it is not economical to write the full address for every part. It makes sense to break down the address and create a special frame for the common segment. In forming an address of a specific part, only the unique component of the address is included, i.e., the bin number. Other segments of the address which are common are determined in the inheritance process consisting of obtaining information while ascending the frame hierarchy.

The frame concept is widely used in constructing problem-oriented KPS. Let us consider as an example some frame types developed for knowledge representation in industrial systems [6]. Here a frame is presented in the format of an oriented graph with marked vertices and arcs. One vertex is assigned to the predicate (functional) symbol, the remaining vertices to arguments of the isolated symbol. A range of permissible values is given for each argument vertex, which permits viewing a given vertex as a slot. Let us note that this frame definition is very close to the concept of a fact in a semantic network. For this reason, some specialists view frames as a special case of specialized networks.

Let us consider specific frames.

Frame-joint describes special types of joints (mechanical, hydraulic, etc.) encountered in industrial systems. A prototype of a frame-joint is shown in Fig. 2.9. It reflects a stereotype situation: "Subject X joins object Y with object Z using method W." The predicate vertex in the figure is marked by F and by the frame name. Arcs carry markers of case relationships: s designates the subject, o the object, and k the relationship "by means of what." Vertices drawn as rectangles with a marker D describe the region of permissible values of a corresponding argument.

Frame-designation serves to describe processes by designating individual elements that are participating in them. Suppose we wish to describe a situation: "Pump P pumps gas G from the source of heat SH to the heat exchanger HE." A prototype of a frame-designation is shown in Fig. 2.10. Symbols u and d denote case relationships "source of action" and "receiver of action," respectively.

Frame-law of operation is used to describe analytical laws of variation of specific parameters in time. A prototype of a frame-law of an operation is shown in Fig. 2.11. It describes the following situation "Compute the value of parameter p at a time instant t by means of function f that has arguments $\alpha_1, \ldots, \alpha_m$." Arc markers denote relationships: v_{1f} is the type of function, τ is the time, arg is the argument of the function, res is the result of the function application.

In systems of knowledge representation based on frames, three main approaches to the organization of information processing and computing are used [1]. In the *first approach*, processing is organized by the user through the utilization of some programming language. More often than not it is LISP. This approach is used in the FRL system. In the *second approach*, a unified information and computing process is used for the frame system based upon a selection of frames that control subsequent computing. Thus, in the KRL system, during the construction of an informational model for a specific situation, one begins by performing a rough selection of base frames and then compares them (using the concept "category") with the constructed frame-samples. The comparison

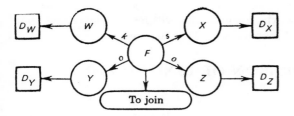

Figure 2.9   Frame-joint.

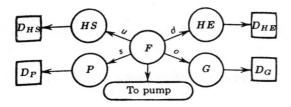

Figure 2.10   Frame-designation.

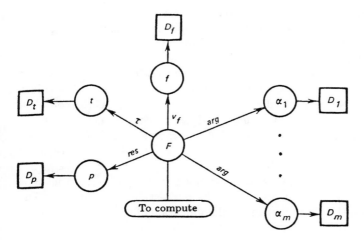

Figure 2.11   Frame-law.

problem may be subdivided into subtasks which permits utilization of a search strategy in the subtask space (see §3.3). In the *third approach*, subclasses of frame systems are established for which specific algorithms are developed that are essentially based on individual properties of these subclasses.

2.5   THE REPRESENTATION OF LINGUISTIC KNOWLEDGE

The theory of grammars by Khomsky is widely accepted as a basis for a formal description of languages. Let us review its fundamentals.

Let V be a nonempty finite set of symbols (elements) designated as a dictionary (alphabet). Symbols may be characters, words, mathematical signs, geometric shapes, etc. A random finite sequence of symbols ω is called a "chain in the dictionary V." The number of symbols in the chain is designated as its length and is denoted by $|\omega|$. Chains are obtained by means of an operation known as conatenation (joining). For instance, $\omega = a_1 a_2 a_3$ where $a_i \epsilon V$. A set of a variety of chains in a dictionary V is called a "language." Languages are defined by a "grammar," i.e., a set of rules that generate all chains of a given language and only them.

There are identifying, generating, and transforming formal grammars.

A formal grammar is "identifying" if for any considered chain it decides whether or not this chain is correct from the point of view of this grammar. A formal grammar is "generating" if it can construct any correct chain. A formal grammar is called "transforming" if for any correctly constructed chain it can construct its transformation in the form of a correct chain.

Let us consider the class of generating grammars. A generating grammar is a quad $G = (T, N, P, S)$ where T is the finite nonempty set of symbols designated as the terminal (main) dictionary; N is the finite nonempty set of symbols designated as the nonterminal (auxiliary) dictionary; P is a finite nonempty set of derivation rules (rules of copying, production); S is the input symbol.

A terminal dictionary is a set of input symbols for constructing chains that are generated by the grammar; the nonterminal dictionary is a set of auxiliary symbols denoting classes of input symbols. A finite set $V = N \cup T$ is the complete dictionary of grammar G. A derivation rule is a finite nonempty set of two-place correlations of the form $\varphi \rightarrow \psi$ where φ and ψ are chains in dictionary V and "$\rightarrow$" is interpreted as "replace for." Chain ω can be directly derived from chain ω' using the rule $\varphi \rightarrow \psi$, if $\omega = a_1 \varphi a_2$, $\omega' = a_1 \psi a_2$, $\{\varphi \rightarrow \psi\} \epsilon P$. A sequence of chains $\varphi = \varphi_0, \varphi_1, \ldots, \varphi_n = \psi$, $n \geq$ is referred to as the derivation of ψ from φ if for each i, $0 \leq i \leq n$, chain φ_{i+1} can be derived directly from φ_i. The derivation length equals the number of times of application of the derivation rules. A derivation of a chain ψ is considered complete if there are no chains that can be derived from ψ.

The input symbol S is an isolated nonterminal symbol that designates a class of language objects to which a given grammar is assigned. The input symbol is sometimes referred to as the target of the grammar or its axiom.

In describing a natural language in the terminology of the theory of formal grammars, terminal symbols are interpreted as words or morphemes, the smallest meaningful language units (roots, suffices, etc.); nonterminal symbols are interpreted as names of classes of words and word formations (subject, predicate, verb group, etc.). A symbol chain is usually interpreted as a natural-language sentence.

Let us consider some examples.

1. Suppose a grammar is given as follows

$$\mathbf{T} = \{a, b\}, \ \mathbf{N} = \{S, A, B\}, \ \mathbf{S} = S$$

$$\mathbf{P} = \{1. \ S \rightarrow aB; \ 2. \ S \rightarrow bA; \ 3. \ A \rightarrow aS; \ 4. \ A \rightarrow bAA$$

$$= \{5. \ A \rightarrow a; \ 6. \ B \rightarrow bS; \ 7. \ B \rightarrow aBB; \ 8. \ B \rightarrow b\}$$

The typical sentence derivations

1. $S \xrightarrow{(1)} aB \xrightarrow{(8)} ab$
2. $S \xrightarrow{(1)} aB \xrightarrow{(6)} abS \xrightarrow{(2)} abbA \xrightarrow{(5)} abba$
3. $S \xrightarrow{(2)} bA \xrightarrow{(5)} ba$
4. $S \xrightarrow{(2)} bA \xrightarrow{(4)} bbAA \xrightarrow{(5)} bbaA \xrightarrow{(5)} bbaa$

A number of a rule used is indicated in parentheses above the arrows.

2. Suppose a grammer is given as follows:

$\mathbf{T}$ = {transports, delivers, container, load, vehicle, powerful, cargo, big};

$\mathbf{N}$ = {P — (predicate), S — (subject), D — (definition), SP — (supplement), SG — (subject group), PG — (predicate group), SN — (sentence)};

$\mathbf{S}$ = {SN — (sentence)}.

$$\mathbf{P} = \begin{cases} 1. & SN \rightarrow (SG)(PG); \\ 2. & SG \rightarrow (D)(S); \\ 3. & PG \rightarrow (P)(SP); \\ 4. & D \rightarrow (big, powerful, cargo); \\ 5. & S \rightarrow (vehicle); \\ 6. & P \rightarrow (transports, delivers); \\ 7. & SP \rightarrow (container, load). \end{cases}$$

Let us construct some language chains:

SN → (SG) (PG) → (D) (S) (PG) → (D) (S) (P) (SP) → BIG (S) (P) (SP)
→ BIG VEHICLE (P) (SP) → BIG VEHICLE TRANSPORTS (SP)
→ BIG VEHICLE TRANSPORTS CONTAINER.

Apparently, the last chain of the derivation is the final chain and is a sentence of natural language. In a similar manner, the chain CARGO VEHICLE DELIVERS LOAD, and a number of other chains, can be derived. Let us point out that in this example syntactic categories are the nonterminal symbols.

A derivation may also be described by a configuration tree that satisfies the following requirements:

1)   Every tree vertex has symbol V as a marker;
2)   The root of the tree, i.e., a vertex that is not entered by any arcs, has the marker S;
3)   If a vertex with a marker D has a least one subordinated vertex, then $D \epsilon N$;
4)   If some n vertices with markers $D_1, D_2, \ldots, D_n$ are directly subordinated to a vertex with the marker D, then at P the rule $D \to D_1 \ldots D_n$ applies.

The configuration tree for the second example is shown in Fig. 2.12. In linguistics, words or sequences that are elements of some other construction are referred to as "components." Therefore, a configuration tree is called a "marker of a component configuration" or a C-marker.

In the theory of formal grammars, there are four types of languages generated by four types of grammars. Grammars are made by imposing progressively stiffening restrictions on the system of rules P.

Grammars of the 0-type are those where there are no restrictions on the derivation rules $\varphi \to \psi$ where φ and ψ may be any chains from V.

Grammars of the type 1 are those that contain rules $\varphi \to \psi$ satisfying the conditions $\varphi = \varphi_1 A \varphi_2$, $\psi = \varphi_1 \omega \varphi_2$, where A is a nonterminal symbol, and φ, φ_1, φ_2, ψ, and ω are chains from the dictionary V. Thus, in grammars of type 1, an individual terminal symbol A transforms into a nonempty chain ω (A may be replaced by ω) only in the context φ_1, φ_2. Grammars of type 1 are called "context" or "context-dependent."

Grammars of type 2 are those where rules of the format $A \to \omega$ are allowed, where $A \epsilon N$, and ω are any nonempty chain from V. Grammars of type 2 are called "noncontext" or "context-free" (CF).

Grammars of type 3 have rules of the format $A \to aB$ or $A \to b$, where A, $B \epsilon N$; a, $b \epsilon T$. Let us note that A, B, a, b are stand-alone symbols (not chains) of the corresponding dictionaries. Languages that are defined by these grammars are called "automatic" or "regular."

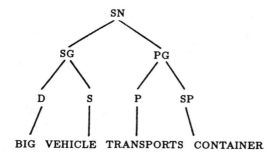

Figure 2.12   Marker of a component structure.

There are other grammar classifications in addition to those considered above. Let us consider the most important ones for the sake of understanding grammars in the future.

"Nontruncating grammars" are those where for any rule $\varphi \to \psi$, the relationship $|\varphi| \leq |\psi|$ is valid. Thus, as a result of utilizing the derivation rule, a chain whose length is not less than the length of the initial chain is obtained.

Suppose the grammar is given in the following manner:

$$\mathbf{T} = \{a, b\}, \ \mathbf{N} = \{S\}, \ \mathbf{S} = S, \ \mathbf{P} = \{S \to aa, \ S \to bb, \ S \to aSa, \ S \to bSb\}$$

It is obvious that for any initial chain the application of the four derivation rules will not permit a chain shorter than the initial one to be obtained. This means that this grammar is nontruncating. The grammar of direct-components has rules of the form $\varphi A \psi \to \varphi \omega \psi$ or $A \to \omega$ where $A \epsilon N$; ω, φ, ψ are arbitrary chains. Apparently the grammar of direct-components (DC-grammar) may be both context-dependent (CD) and context-free (CF). In the DC-grammar, only one symbol may be replaced during any one step. Therefore a derivation in the DC-grammar is convenient to represent in the form of a component tree (C-marker).

DC-grammars and nontruncating grammars equivalent to them in terms of generating power are sufficient (although not always convenient) for describing a sentence in natural languages. However, as with any generating grammar, the DC-grammar contains only rules of formation, not rules of transformation of language expressions. A complete command of a language implies the ability not only to construct tame (correct) sentences, but also to make transitions from one tamely-constructed sentence to another that may be synonymous to the first sentence or that may differ from it in some specific way. Examples of such transitions are transformations of an assertive sentence into a question, an active form into a passive, etc. In other words, a generating grammar operates only with the syntax of a language and not its semantics. For describing the semantics, it is necessary to introduce rules of transition from an expression to another one which carries the same information. Grammars that utilize such rules are called transformative.

At the basis of transformative grammars is the assumption that the generation of a sentence by a human is performed in two phases. During the first phase, a relatively simple syntactic structure is created (referred to as profound) by means of some grammar. During the second phase, this structure by means of sequential application of transformation rules is converted into a more complex syntactic structure designated as superficial and corresponding to a specific sentence. The initial DC-grammar is called the base grammar and transformation rules are called transformations. Formally, a profound structure is a C-marker generated by categorizing rules, i.e., derivation rules that contain no terminal symbols. Chains generated by them contain only nonterminal symbols designating syntactic language categories. A substitution of terminal symbols into the generated sentences is realized by means of special rules for inclusion.

A transformative grammar may be formally presented in a two-component format. The first component is the DC-grammar; the second, the so-called T-component, is formed on the basis of transformation rules $p_1, \ldots, p_n \to l$, where $n \geq 1$ and $p_1, \ldots, p_n, l$ are not symbol chains but C-markers. Thus, a derivation in the transformative grammar is the transformation of a set of C-markers into a new C-marker. The result of transformative transformations is a superficial sentence structure.

Direct utilization of transformative grammars in the synthesis and analysis of natural language sentences represents a number of difficulties. Thus, in generating a sentence, a superficial structure, as was shown above, is obtained by means of transforming a profound structure. In so doing, the question of the tameness of the superficial structure may be uniquely answered only by means of a corresponding grammar that is different from the base grammar. The construction of such a grammar is difficult. There are also both theoretical and practical complications. The desire to overcome them led to the development of analysis models furthering the ideas of transformative grammars. The expanded transition network, for instance, that described in §4.1, falls in this category.

2.6   LANGUAGES FOR KNOWLEDGE REPRESENTATION

Models of knowledge representation may be realized by means of software in widely used universal programming languages FORTRAN, ALGOL, etc. With their aid, one can interpret any knowledge representation model, since in these languages there is a distinct declarative part (description) and a procedural part (sequence of operators implementing algorithms). However, the great complexity and laboriousness of such interpretations makes the development of practically useful systems of artificial intelligence possible.

The desire to obtain an efficient software realization of knowledge representation models resulted in the development of a large number of knowledge representation languages (KRL) ranging from simple, dedicated for solving special isolated problems, to powerful universal languages.

Common properties of high-level KRL include:

- availability of means for describing more complex types of data and control procedures than those in universal programming languages;

- availability of built-in mechanisms of representation, search, and processing of information;

- availability of tools for constructing deductive algorithms.

Knowledge representation languages may be conditionally divided into three groups [7]:

1. Languages for processing symbolic information. Typical representatives of this group include LISP, REFAL, SNOBOL.

2. Languages designed to seek solutions in a state space and proofs of theorems (see §3.2 and 3.4). They include, for instance, PLANNER, QLISP, QA-4, and others.

3. General purpose knowledge representation languages. Representatives of this group are KRL and FRL. They are sometimes referred to as second-generation languages, since chronologically they evolved after those mentioned above. Accordingly, they use the latest artificial intelligence ideas.

Let us examine some features of the above groups.

Languages for processing symbolic information are the most adequate for programming a large class of artificial intelligence problems. This is due to the fact that the basis of all forms of intellectual activity is the ability to process symbols, to transform their sequences into complex structures, and to perform various operations on them.

The most widely accepted language for processing symbolic information is LISP. Let us examine some aspects of this language.

The basic information units in LISP are atoms and lists. An atom is any sequence of alphanumeric symbols and special signs bounded by some limitations. Atoms may be used for representing syntactic indices, semantic markers, names, relationships, operations, and properties, as well as variables. A list is an ordered aggregate of elements that may be atoms or other lists. Lists are constructed in the form of balanced (in terms of parentheses) structures that permit arbitrary insertiveness. In LISP, an expression means an atom or a list. Expressions that have values are constants, variables, function names, and functions. The expression (6, 8, SPM, B) represents a list of four elements each of which is an atom; the expression (PUT(THE PEN)(IN(THE TABLE))) is a three-element list. The first element is an atom, the second a list consisting of two atoms, and the third a two-element list with the second element in the format of a list. The ability to include lists into other lists by means of parentheses permits the construction of tree-like structures of data and perform their transformations.

Special values may be assigned to individual list positions; for instance, (#SUP-PORT AB) is a three-element list with the first position "#" being a relationship sign, the second, SUPPORT, is the name of the relationship, while the entire expression denotes a two-place relationship "B is supported by A."

LISP has a means for interpreting lists designated as computation. The first element of a list is interpreted as the function name, while the remainder of the list is viewed as a set of arguments of the function. Let us examine a list computation in case of arithmetic functions. A list (PLUS 3 4) produces a value $7 = 3 + 4$. A list (TIMES 9 3) produces a value $27 = 9 \times 3$. A list (MAX 2 8 6) produces a value of 8, i.e., the maximum out of three values of the arguments. A list (PLUS(TIMES 2 5)(QUOTIENT 6 2)) is interpreted as $(2 \times 5) + (6 : 2)$. Functions TIMES, PLUS, QUOTIENT, and others are built-in LISP functions. Besides built-in functions, the user may construct and enter his own functions. For instance, to describe the expressions of the format $x^2 + 6xy + 8$, the expression (DEFINE(F X Y)(PLUS)(SQUARE X)(TIMES 6 X Y)8) is constructed. The function DEFINE is of the class of noncomputable functions and is used to define a new function $(F, X Y)$, where F is the name of the new function, and X and Y are its arguments. When the system encounters the expression $(F\ 5\ 10)$ in the program text, it will let $X = 5$, $Y = 10$ and will compute function F in accordance with the rule written in the list after the list $(F\ X\ Y)$. The function SQUARE is also not built-in and thus must be defined: (DEFINE(SQUARE X)(TIMES X X)).

For assigning a value to a variable, one uses function SETQ. For instance, (SETQ R 4) — variable R will assume the value 4. (SETQ B(PLUS 3 5)) — variable B will assume the value 8.

For processing lists, in LISP there is a number of elementary functions such as CAR, CDR, CONS, and others. Function CAR chooses the first element of a list, an argument, as a value; function CDR chooses all list elements except the first. For instance, (CAR(ABC)) is A, (CDR(ABC)) is (BC). It is obvious that a combination of these functions permits the selection of any atom in a list. Suppose that it is necessary to select the second atom of the list (ABC). This can be done by means of the expression (CAR(CAR'(ABC)). Here an apostrophe is a divider indicating the beginning of a processed list.

While functions CAR and CDR are used for separating lists, function CONS combines lists. Suppose there are two lists, (A) and (BC). The expression (CONS $A(BC)$) produces list (ABC), i.e., the atom located in the first argument place becomes the first element of a new list.

In LISP there are several base predicate functions that assume values T (true) and NIL (false). These include ATOM, NULL, EQ, and others. The function ATOM verifies whether its argument is an atom, while function NULL verifies whether its argument is an empty list. Function EQ verifies equality of two arguments. A combination of base predicate functions permits the construction of functions of any complexity.

The noncomputing function DEFINE may be used for entering list processing functions. Suppose there are two symbolic variables: APPLICANT and STUDENTS. Let us define values of these variables:

(SETQ STUDENTS (PETROV IVANOV))
(SETQ APPLICANT (SIDOROV VASILEV))

Let us assume that the list of students may be expanded at the expense of elements of the applicant list. Then, if VASILEV becomes a student, his name should be removed from the applicant list and entered into the student list. Let us construct a function INCLUDE, that performs such operations:

(DEFINE (INCLUDE NAME)(SETQ APPLICANTS
(DELETE NAME APPLICANTS))
(SETQ STUDENTS (CONS NAME STUDENTS)))

The function DELETE deletes an element located in the first argument place whose name is in the second argument place from the list. Having at our disposal the function INCLUDE, the transfer from the list APPLICANTS to the list STUDENTS is performed by the expression (INCLUDE VASILEV).

Here we examined the simplest concepts of the LISP language. At this time there are numerous dialects of LISP and voluminous libraries of LISP programs. This makes this language at least as efficient as any other and even superior in terms of convenience of interfacing and service capabilities.

Languages oriented toward the solution of search problems and theorem proving (languages of the second group) are characterized by sufficiently simple and well-defined, but complicated information processing methods. Therefore, in languages of this group, the main attention is being paid to the efficiency of the mechanism of program control. The most widely accepted language is PLANNER [8], which combines the LISP capabilities with a developed method for comparison of samples, search with retrieval, and procedure calling in accordance with a sample.

Sample comparison permits the construction of programs of data analysis in a simpler and more compact form than that in the case of LISP alone. A syntactic sample is a templet that is superimposed upon an analyzed expression in order to determine whether it has the required structure. The process of this superimposition is referred to as comparison. The ability to set up procedures of the associative search and call up procedures on the basis of their samples is its most important feature.

The mechanism of a search and retrieval is a method for running a program by performing trial steps (solutions). If they lead to a dead end, the program abandons them. This mechanism permits the user to easily set up various sorting algorithms, which is quite important since sorting algorithms characterize search problems in the state space.

The language PLANNER offers tools for describing an object domain in the form of a database consisting of random LISP expressions that describe some specific fact. For describing laws and relationships among individual facts, one uses the so-called theorems which are procedures that are called up on the basis of their samples. A theorem consists of a sample and a body. When a new expression is added to the database, the program automatically calls up all theorems whose samples may be compared to this expression, and the bodies of these theorems are executed. The execution of a theorem body may generate expressions that may call up other expressions. This mechanism, for instance, may be used to realize search procedures by means of a method for the reduction of a problem to subproblems.

"General purpose" knowledge representation languages (languages of the third group) are characterized by these features [7]: data is presented in the form of a multilayer frame structure; an abstract model of an object domain is presented in the form of a hierarchically-organized set of concepts, while a specific model of a situation is presented in the form of a combination of mutually linked copies of these concepts; mechanisms of semantic (developed on the basis of syntactic) comparison are utilized. Languages in this group include KRL and FRL mentioned earlier. Besides these two, there is KL-ONE, which has a number of common features with KRL, language Φ that makes extensive use of frame and other concepts.

Most languages of the second and third groups are either expanded versions of LISP or actively use expanded procedures written in LISP. In recent years another-than-LISP language has been gaining acceptance in the artificial intelligence area, that is, PROLOG. It is a high-level language based on mathematical logic. Its characteristic feature is the availability of several equal-right semantics. Text in PROLOG may be treated as a declarative description of certain properties and as a description of procedures. The main unit of a PROLOG program is a description of a relationship, while a program represents a set of descriptions of relationships and a database containing facts which are specimens of correlations. All requests in the language are formulated in the terminology of entered or built-in correlations. There are three types of requests: verify validity of an expression; output all objects satisfying a description; add a fact.

Program representation in logical terms is convenient when the number of the utilized correlations is not very large. If a hundred or more relationships are required for describing a problem, then a PROLOG program becomes quite cumbersome and complex for understanding, which complicates its modifications and reduces its reliability. Therefore, it is difficult to visualize PROLOG as a dominant language for knowledge representation. It appears that it will take its place alongside LISP and its dialects.

THREE
METHODS OF SOLVING PROBLEMS

The process of problem solving as a rule contains two phases: problem representation and a search (a sorting procedure). The successful solution of a problem depends to a large degree on the form of its presentation. Forms or methods of problem presentation may vary and depend both on the nature of the problem and on its solver. For instance, in integrating, a person attempts, by means of transformations, to represent the integral in such a way as to be able to use tabular integrals. In selecting the shortest path, a person may use graphic representations of possible travel paths and evaluations of their distances among intermediate points. However, the nature of human thinking is such that the route and the form of problem representation that is being used are not always conscientiously understood by a person. Thus, a chess player cannot rigorously explain the form of representation of a chess position that permits him to avoid sorting all possible variations while searching for the next move. Therefore, the phase of problem representation frequently is not clearly understood by man. Nevertheless, the importance of this phase is realized as soon as one attempts to develop a software-realizable algorithm of a problem solution.

3.1 METHODS OF PROBLEM PRESENTATION

The search for a problem presentation format that would be convenient for its machine solution is a creative process that is difficult to formalize. One can point out the following more popular formats: representation in the state space; representation by reducing the problem to subproblems; representation in the form of a theorem; a combinatorial representation.

Representation of problems in the state space. A complete representation of a problem in the state space includes the description of all states or of only the initial states, the definition of operators that transform some states into others, and the definition of a target state. Different forms of a description of problem states are possible. For instance, one may use lines, vectors, matrices, and graphs. In selecting a form of description of states, it should be ensured that the use of an operator for transforming states is sufficiently simple.

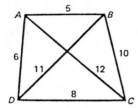

Figure 3.1   Route diagram.

Operators may be specified by means of a table relating each "input" state with some "output" state. For large problems, such an operator definition is not practical. In describing states in the form of a line (vector), it is convenient to define the operator in the form of a rule for rewriting.

The procedure of a solution search in the state space involves the determination of a sequence of operators that will transform the initial state into the target state, i.e., the state satisfying the desired goal. Problem solution will be this sequence of operators.

It follows from this that representation of a problem in the state space is determined by a combination of three components: a triad (S_0, F, G), where S_0 is a set of initial states; F is a set of operators that transform one state into another (e.g., the state space onto itself), G is a set of target states.

Problem of route selection or the problem of a travelling salesman. A transportation robot should lay out its route in such a manner that it stops at each one of n given points exactly once and returns to the originating point. In addition, if there are several such routes, it is desirable to select the shortest one. Figure 3.1 depicts a diagram showing the distribution of points A, B, C, and D, routes, and distances.

States in this problem may be given by a line denoting the list of points visited at the moment. Thus, if the robot begins its travel from point A, then the initial state will be represented by a line consisting of a single character A. A point cannot be mentioned more than once in any given line. An exception is a line at the end of which point A is repeated after all other points are enumerated (mentioned).

Operators are transformations corresponding to the robot's decision to travel from a given point to some other point. A target state, if there is no restriction on minimizing the length of the route, is any state whose description begins and ends with A and contains the names of all other points. Such a state, for instance, is $ABCDA$.

The problem of object manipulation. Suppose a state or, as it is commonly referred to, a problem universe, includes some number of platforms with cubes (in a true-life situation, instead of cubes, there may be parts which one wants to assemble into a finished product). There is also a robot-manipulator that carries cubes from one platform to another and places the cubes on top of one another. A current state of such a universe may be viewed as a current positioning of cubes and the location of the manipulator. A combination of all possible states comprises a set **S**. It is obvious that the methods of moving cubes by the robot, i.e., transitions from one state to another, must obey certain rules. Thus, it may be specified that only a certain number of cubes is placed on a specific platform. In picking up cubes to be carried away, it is natural to select those that are at the top, etc. On the basis of such restrictions, a set **F** of permissible operators of universe transformation which, in effect, are a combination of permissible robot movements, are constructed. The desired positioning of cubes at the platforms is the target state of the universe. A solution consists of a sequence of robot actions (operator chain $F_1, \ldots, F_n$) required to change the initial state (the initial state of the universe) into a desired one, i.e., into the target state. This problem is often referred to as the "problem of scene transformation."

Planning the operation of a flexible manufacturing cell. Let us examine a flexible manufacturing cell consisting of several machining centers (that will be different in a general case), an automated guided vehicle (AGV), and an automated warehouse. The cell has to process N parts of m different types during a time period that does not exceed T. The initial state of the system all blanks required to machine parts are in the warehouse, the AGV and the machining centers are in some specific state (the initial states of the system). All blanks are processed as required at the appropriate machining centers (the target state). The current state of the system is determined by the number and type of completed parts and states of the machining centers and the AGV, while the state of a machining center is determined by the type of a part being machined (waiting for a part and downtime of a machining center are also among its specific states). The state of the AGV depends on its location, on which part it is carrying at a given time, and on whether it is in motion or waiting to be called. The set of operators **F** is prepared by taking into the account functional route requirements, the machining centers setup, and AGV capabilities. Every displacement of a part from one machining center to another is, obviously, a change in the state of the system. The problem solution will be given by an operator sequence $F_1, \ldots, F_n$ that arranges the system state changes (movements of parts between the machining centers) in such a way as to ensure that all parts are completed in time interval that does not exceed T.

The graph. In describing a state space and the methods of the search for solutions by applying artificial intelligence, a graph is widely used, especially its variation referred to as a tree.

A graph (or a geometric graph) is a geometric configuration of **V** points that are interrelated with a set **E** of continuous, non-self-intersecting curves [2]. Points of set **V** are called "vertices" and curves of set **F** are called "edges." Common points of edges may only be vertices (i.e., points of set **V**). If all edges have a specific direction (denoted by an arrow), then the graph is called an "oriented graph" or "orgraph" while its edges are called "arcs." If there are given two graphs G_1 and G_2, and sets of vertices and curves of graph G_2 are subsets of sets of vertices and curves of G_1, respectively, then graph G_2 is designated as a "subgraph" of graph G_1, while graph G_1 is designated as a "supergraph" of graph G_2. A sequence of different arcs $d_1, \ldots, d_n$, for which the terminal vertex v_i of arc d_i is the starting vertex of arc d_{i+1} $(i = 1, \ldots, n - 1)$, is referred to as a "path" from vertex v_0 to vertex v_n (v_0 is the starting vertex of arc d_1, v_n is the terminal vertex of arc d_n).

A "tree" is a graph where each vertex is entered by only one arc, except the vertex known as the tree root. No arc enters a root. Thus, in a tree, every vertex with the exception of the root is the end of only one arc and the beginning of one or more other arcs, while a root can only be the beginning of one or several arcs. All arcs originating from a vertex are called "branches" of the vertex. If v is some vertex of a tree and vertices v_i $(i = 1, \ldots, m)$ are ends of arcs originating from the vertex v, then we will say that vertices v_i are generating by vertex v or that vertex v generates vertices v_i. Sometimes vertex v (relative to vertices v_i) is called the "parent," while vertices v_i (relative to vertex v) are "filial" vertices or "receivers."

Let us introduce a concept of "level." Let us agree that a root is at the zero level; vertices generated by the root are at the first level; vertices generated by vertices of the k-th level are at the $(k + 1)$ level.

Representation that reduces a problem to subproblems. Such a representation provides a subdivision of an initial problem into a set of subproblems, the individual solutions of which produce the solution of the initial problem. Each subproblem may, in turn, be broken down into subproblems. The number of subdivision levels theoretically is unlimited. In practice, a subdivision continues until a set of problems (subproblem) is derived at a lower level, such that a method of their solution is known. Such problems will be designated as "elementary."

There are two types of mutual relations among subproblems: AND-structures and AND-OR-structures. In structures of the AND type, all subproblems must be solved to obtain the solution of the main problem. In AND-OR structures, problems are broken down into groups. Within the groups, they are linked by AND relationships, while the groups are linked by OR relationships. In this case, to obtain the solution of the main problem, it is sufficient to solve all subproblems within any one group.

To describe the representation of problem reduction to subproblems, one may utilize a graph called a "graph of problem reduction." Here vertices will correspond to solutions while arcs will correspond to operators of problem reduction. The initial problem corresponds to the root of the tree, problems generated directly by the initial problem correspond to vertices of the first level, problems generated by the first level to the vertices of the second level, etc. Let us assume that problem A may be solved if one solves problems B and C or D; problem B will be solved if one solves problems E and F; problem C will be solved if problem

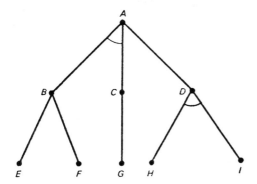

Figure 3.2   Problem reduction tree.

G is solved; problem D will be solved if H and I are solved. A graph of such a representation of problem A is shown in Fig. 3.2. Vertices B and C are linked in the sense that, to obtain the solution for problem A, it is necessary to solve both problems B and C. Vertices H and I are also related in this sense. To denote a relationship among some vertices, arcs entering these vertices are connected by a special curve. If there are related vertices (arcs), then it is common practice (adding, if necessary, auxiliary vertices) to transform the reduction tree in such a way as to ensure that each group of related vertices has a separate parent vertex. In Fig. 3.2, related vertices B and C have a common (with D) parent vertex A. After transforming this tree in accordance with the indicated requirement (introduction of an additional vertex K), we obtain a tree shown in Fig. 3.3, in which each group of related vertices has a separate parent vertex. Below we will be dealing with trees of problem reduction of this type only.

Structurally a graph (tree) of problem reduction differs from a graph (tree) of the state in that it has linked arcs. Linked vertices are also called AND-vertices, while nonlinked are called OR-vertices, and a graph containing AND- and OR-vertices is called a graph of the type AND-OR. Vertices corresponding to elementary problems are referred to as "terminal Z." Vertices that do not have any filial vertices and are not terminal will be referred to as "dead-end" vertices. Dead-end vertices are associated with problems that are insolvable in the form of a given representation, i.e., they cannot be solved. Vertices corresponding to solvable problems are referred to as "solvable," while vertices that are associated with insolvable problems are referred to as "insolvable." Accordingly, terminal vertices are solvable, while dead-end vertices are insolvable. A vertex that is neither terminal nor a dead-end will be solvable when and only when all of its filial vertices are solvable, if they are linked, or when at least one filial vertex is solvable if they are nonlinked. It is obvious that problem A (Fig. 3.3) is solvable in that (and only in that) case when vertices H and I are terminal or when vertices G and F or E are terminal.

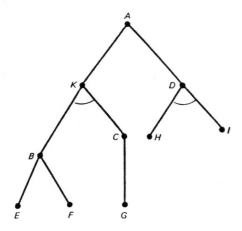

Figure 3.3 Transformed reduction tree.

A graph of problem reduction may be specified in an explicit form, as in the example just considered. More often than not, just like the state graph, it is given implicitly in the form of a description of the initial problem and reduction operators.

Representation in the form of theorems. Many problems (of the logic type and others) may be formulated as theorems that are subject to proof. Among theorem proofs one can list puzzles, game problems, as well as a number of practical problems, such as decision making and operations planning.

In representing a problem in the form of a theorem, premises, i.e., a set of known true statements (axioms), are determined or given. After this, a theorem is formulated whose proof permits the initial problem solution to be obtained. The solution search for such a representation is performed in the following manner. By combining two or more axioms, statements that may be obtained from premises are derived. Next, whether the derived set contains the theorem or its negation is verified. If it does, then the theorem is proven or, respectively, rejected. If neither the theorem nor its negation is contained in the set, then the derived set is added to the premises and the procedure is repeated. This method of a search for a solution (proof) is referred to as a method of a complete sorting. If the theorem to be proven or its negation may be derived on the basis of initial premises, then by using the method of a complete sorting one can always obtain the desired proof using a finite number of steps. In practice, the utilization of this method is limited, since during a complete sorting an excessive amount of memory and a long time (to arrive at the solution) may be required.

Combination representation. In general, it is not mandatory to use a specific representation. A combined (e.g., mixed) representation may be more efficient in solving specific problems. For instance, if the problem has an obvious hierarchial structure and can be easily broken down into a number of subproblems, then it is natural to first reduce the problems to subproblems, and for the representation of subproblems to use different forms of description in the state space or representation as a theorem.

3.2 SOLUTION SEARCH IN THE STATE SPACE

The solution search in the state space boils down to the determination of an operator sequence that transforms initial states into the target states. If there is more than one such operator sequence and a criterion is given for determining an optimum sequence, then the search entails finding the optimum sequence, i.e., an operator sequence that ensures an optimum of a given optimization criterion. It is convenient to examine the methods for the solution search in the state space using a tree (graph) of states. Using a state tree, the solution search is reduced to the determination of a path (optimum, if there is an optimization criterion) from the root of the tree to the target vertex, i.e., to a vertex that corresponds to the target state. The solution search may be visually demonstrated using a state tree when there is only one initial state. For this reason, let us first examine those problems that are defined by a triad $(\mathbf{S}_0, \mathbf{F}, \mathbf{G})$, where set $\mathbf{S}_0$ of initial states consists of one element. It should be remembered that in this triad, $\mathbf{F}$ is a set of operators transforming a state space onto itself; $\mathbf{G}$ is a set of target states.

To construct a state tree, one should, using $\mathbf{F}$ operators applicable to the tree root (the initial state), construct vertices of the first level. Next, using $\mathbf{F}$ operators applicable to the first level vertices, vertices of the second level should be constructed, etc. The process of operator application to a vertex of the tree for constructing all of its filial vertices is known as the "uncovering" of a vertex. Therefore, operators of universe transformation (operators from $\mathbf{F}$) are interpreted as rules of uncovering vertices. The application of uncovering rules to the starting vertex generates a combination of filial vertices. Each of them corresponds to a certain state of the universe, to which it may make a transition from the initial state. Arcs that are linking the initial vertex with a filial vertex are identified as the corresponding transformation operators. For all filial vertices, one verifies whether they are target vertices, i.e., whether they correspond to the target states. If a target vertex is not uncovered, then the next step of vertices generation is performed by applying the uncovering rules to each vertex generated during a preceding step. The derived set of vertices is also analyzed for the presence of the target vertex. The procedure continues until the target vertex is located.

Let us examine a procedure of the construction of a state graph using an example of route determination by an AGV which, after departing from the warehouse (point A), should stop at all machining centers (points B, C, D) and return to the warehouse. The AGV cannot visit any machining center more than once. The routes and distances are shown in Fig. 3.1.

In this problem, we will denote a state by a word (a set of characters) made up from characters that correspond to the names of points that the robot visited up to that time and arranged in the order in which the points were visited. Then it is obvious that state A will be the initial state, while target states will be those that are denoted by words beginning and ending with the character A and containing the names of all other points. States which are described by words containing repetitions of characters, are not permissible, with the exception of the target states.

Operators (or uncovering rules) in this example correspond to the selection of this or another route. Since from any point any other three points can be reached, the set consists of 12 operators. No more than three operators can be applied to any given state.

In the graph of states of this problem (Fig. 3.4), state A corresponds to the initial vertex (tree root). The tree root generates three filial vertices corresponding to states AB, AC, and AD. Each of the vertices generated by the root generates two vertices, and each of the second and third level vertices generates one vertex. Distances that the robot travels during transitions from one state to another are marked at the arcs. At the terminal vertices that correspond to target states, in addition to states, distances that the robot travels from the initial state to each target state are shown.

The solution search is of the iterative nature and the number of iterations and vertices uncovered before a target vertex depends in large measure upon the order (sequence) in which the vertices are uncovered. The order of vertex uncovering is called the "search strategy."

There are two types of search strategies: a "blind" sorting and an ordered sorting of vertices-candidates to be uncovered. The blind sorting is characterized by the fact that the location of target vertices or their proximity do not effect the order of uncovering. There are several blind-sorting algorithms. Let us examine three of the most common: the algorithm of a complete sorting, the algorithm of equal prices, and the algorithm of an in-depth sorting.

The algorithm of a complete sorting. Vertices are uncovered in the order in which they were generated. The initial vertex is uncovered first. A check is made whether there is a target vertex among the generated vertices. If there is, then the search is completed. If not, then the first vertex among the generated vertices is uncovered and checked to see if it is a target vertex. Then the second generated vertex is uncovered, etc.

For a structured record of algorithms for performing a search in the state space, let us introduce the concepts of lists of open (OPN) and closed (CSD) vertices. The "list of closed vertices" is a list that holds identifiers of uncovered vertices and the identifier of a vertex to be uncovered at a given instant of time. Vertices from the CSD list, except the last vertex, cannot be uncovered. The "list of open vertices" is a list containing vertices that can and will be uncovered. Search strategies differ by the rules for arranging vertices in the OPN list and by the selection of the next vertex to be uncovered.

In a structured form, an algorithm of a complete sorting may be presented as follows:

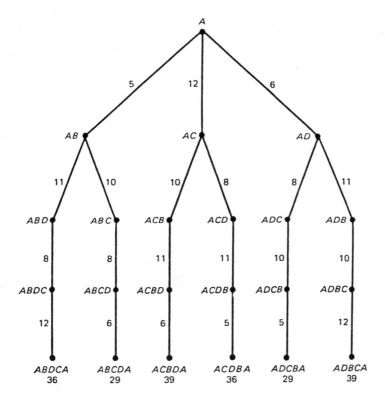

Figure 3.4 State graph of the route-selection problem.

1. Place the initial vertex into the OPN list.

2. If the OPN list is empty, then issue a signal that the search failed, otherwise go to the next step.

3. Take the first vertex from the OPN list and transfer it to the CSD list; then assign an identifier v to this vertex.

4. Uncover vertex v. Place all filial nonrepeating (e.g., not encountered in the CSD list) vertices at the end of the OPN list and construct pointers leading from them to vertex v. If vertex v does not have any filial vertices or has only repeating filial vertices, then go to step **2**.

5. Verify whether any of the filial vertices is a target vertex. If it is, then output the solution; otherwise go to step **2**.

In this algorithm it is assumed that the starting vertex (root) cannot be a target vertex. Undoubtedly, this algorithm permits an optimum (minimal number of arcs) solution to be found. For instance, the utilization of this algorithm to solve the problem of the selection of a shortest route essentially reduces to the construction of a graph that includes distances (Fig. 3.4). Having this graph and comparing distances of different routes leading to the target states, it is possible to select optimum routes. As it follows from the state graph, there are two such routes: routes leading to the target states $ABCDA$ and $ADCBA$.

The algorithm of equal prices. Suppose that each arc is assigned some cost function c_{ij}. It is necessary to determine the path of least cost. The uncovering of vertices is done in the order of increasing cost. Let us denote the cost of uncovering some vertex v by $g(v)$ and describe the equal-price algorithm.

1. Place the initial vertex s_0 into the OPN list. Let $g(s_0) = 0$.

2. If the list OPN is empty, issue a signal that the search failed; otherwise, proceed with the next step.

3. Take from the OPN list a vertex for which $g(v)$ has the minimum value and place it in the CSD list. Assign to this vertex the identifier v (if there are more minimum-cost vertices than one, any one may be chosen).

4. If v is a target vertex, output the solution; otherwise go to the next step.

5. Uncover vertex v. For each filial vertex v_i calculate cost $g(v_i)$ using the following equation: $g(v_i) = g(v) + c(v, v_i)$. Place these vertices together with the corresponding $g(v_i)$ into the OPN list and construct pointers leading from them to vertex v. If vertex v does not have any filial vertices, immediately proceed to step **2**.

6. Go to step **2**.

The equal-price algorithm is a generalized algorithm of a complete sort and may be reduced to it under the condition that costs of uncovering all vertices are equal. In applying this algorithm for solving the problem of choosing a route, a lesser number of vertices is uncovered. The graph of the solution is shown in Fig. 3.5, where arcs are marked with "costs" $c(v, v_i)$, and vertices have, in addition to state markers, markers with values of function $g(v_i)$, and also with (in parentheses) numbers indicating the order of their uncovering.

The algorithm of an in-depth sorting. Let us define the depth of a vertex by a number equal to the number of its level. In the method of in-depth sorting, a vertex that has the largest depth is always undercovered. Since several vertices may have equal maximum depth, it is assumed that there is a rule for selecting one of them. In addition, usually due to some considerations, a limiting depth is specified; vertices that have depth equal to the boundary depth are not uncovered. Thus, the method of an in-depth sort may be defined as such a sorting method when one always uncovers a vertex at a maximum depth which is less than the boundary depth.

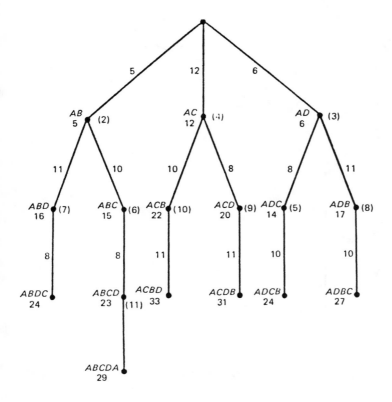

Figure 3.5 Solution graph of the route-selection problem using the equal-price algorithm.

Let us examine the algorithm of an in-depth sort in its structured form.

1. Place the initial vertex into the OPN list.

2. If the OPN list is empty, signal failure, otherwise go to step **3**.

3. Take the first vertex from the OPN list and transfer it to the CSD list. Assign an identifier v to this vertex.

4. If depth of vertex v equals the boundary depth, go to step **2**, otherwise, go to step **5**.

5. Uncover vertex v. Place all filial vertices at the beginning of the OPN list and construct pointers leading from them to vertex v. If v does not have any filial vertices, go to step **2**.

6. If one of these vertices is a target vertex, then output the solution obtained by a retrospective scan in accordance with the pointers, otherwise go to step **2**.

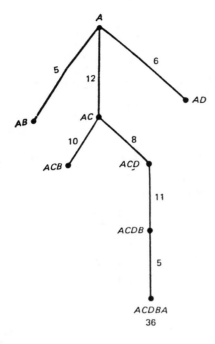

Figure 3.6   Solution graph of the route-selection problem using the in-depth sorting algorithm.

One possible graph for solving the route selection problem by means of the algorithm of an in-depth sorting is shown in Fig. 3.6. When using this algorithm, the solution depends on the strategy of vertex ordering in the OPN list. In the solution shown, a nonoptimum route is determined.

In all the examined sorting algorithms, it is assumed that there is only one initial vertex. If there are several, then these algorithms change only at step **1**: all initial vertices are entered into the OPN list in step **1**.

The advantage of the blind-sorting method is, first, the simplicity of its algorithmic realization, and second, the fact that the solution is always obtained (if it exists). The disadvantage of these methods is the sharp increase in the number of vertices that must be uncovered during the solution search with an increase in the problem size. This substantially narrows the range of practical problems that may be solved by the blind-sorting method.

Algorithms of the orderly sorting. For the majority of practical problems, it is possible to formulate empiric rules permitting one to reduce the sorting volume. These rules utilize specific information about the problem at hand that is being formulated on the basis of the experience, intuition, and common sense of the researcher. The information of this kind is often called "heuristic" and algorithms based upon it are also called "heuristic."

The main idea behind heuristic algorithms is in ordering the list of uncovered vertices in accordance with a standard that evaluates the "potential" of a vertex or path at which a given vertex is located. Such a standard is referred to as the "estimating function." For a recurrent uncovering, a vertex having a minimum value of the estimating function is selected from the OPN list. After each uncovering step, a reordering of vertices in the list is performed in accordance with the values of the estimating function. Such a procedure is called the "algorithm of an orderly sorting."

There are various philosophies of constructing estimating functions. Let us examine one that is used most often [1].

Let $g(v)$ be the cost of the shortest (optimum) path from any initial vertex $s_0 \in S_0$ to some vertex v. The cost of the shortest path from vertex v to the nearest target vertex $g_0 \in G$ will be denoted by $h(v)$. It is obvious that the function $f(v) = g(v) + h(v)$ expresses the cost of the shortest (optimum) path from the initial vertex to the target vertex if it passes through vertex v.

Let us introduce the following estimating functions in this analysis: $\hat{g}(v)$ is the estimate of cost of the shortest path from the initial vertex to vertex v; $\hat{h}(v)$ is the estimate of cost of the shortest path from vertex v to the nearest target vertex. Then the function

$$\hat{f}(v) = \hat{g}(v) + \hat{h}(v) \qquad (3.1)$$

will be the estimating function $f(v)$, i.e., it is an estimate of the cost of the shortest path from the initial vertex to the target vertex under the condition that it passes through vertex v.

It makes sense to choose as an estimating function $\hat{g}(v)$ the actual cost of a path established by the sorting algorithm at a given moment from the initial vertex to vertex v. Let us note that if the graph of states does not have a tree structure, then the value of $\hat{g}(v)$ may vary in the process of uncovering vertices. Let us explain this by means of an example (Fig. 3.7). Suppose that in step 1 of the search there is an uncovered vertex s_0 and there are generated vertices v_i and

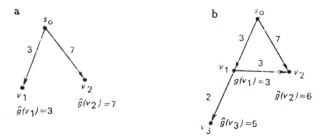

Figure 3.7   State graph that does not have a tree structure.

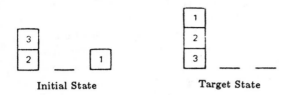

Figure 3.8   Cube location on platforms.

v_2 with cost values $\hat{g}(v_1) = 3$ and $\hat{g}(v_2) = 7$ (Fig. 3.7a). In step **2**, vertex v_1 is uncovered and vertex v_3 and again vertex v_2 are generated. Here the path with the cost $\hat{g}(v_1) = 3$ (Fig. 3.7b) leads from v_1 to v_2. It is obvious that the cost of the "shortest" path from s_0 to v_2 will be $\hat{g}(v_2) = 6$.

This is as follows from the definition and from the examined example, $\hat{g}(v) \geq g(v)$. This estimating function can be easily computed during the process of algorithm application.

The definition of the estimating function $\hat{h}(v)$ is a more complex problem. In its construction, one utilizes any heuristic information about the problem begin solved and therefore $\hat{h}(v)$ is called the "heuristic function." It is obvious that if $\hat{h}(v) = 0$, then the sorting algorithm is reducing to the algorithm of equal prices, which means a complete absence of heuristic information.

There are no constructive recommendations regarding the definition of this function, so let us examine its meaning by means of an example. Let us define the problem of scene transformation (the problem of manipulating objects) in the following manner (see §3.1). Three numbered cubes are located at three platforms. A manipulating robot may move one cube from one platform at a time. Cubes may be placed on top on one another. To move a cube, one may take only the top cube. The initial and target cube configurations are given. The problem is to determine a plan for cube relocation that will permit the transformation of the initial state into the target state with a minimum number of steps (a step is one relocation of a cube). Figure 3.8 depicts the initial and target configuration of the cubes on the platforms.

As a heuristic function $\hat{h}(v)$, let us take the sum of cubes that are not located at their places and the total number of cubes that prevent the transfer of each of them to its own place. The "obstructing" cubes are those that are located on top of a cube that has to be relocated; they are cubes that occupy "someone else's" place in the target configuration, i.e., places that should receive cubes located in wrong places.

The graph of the problem solution using the algorithm of an orderly sorting for the selected heuristic function is shown in Fig. 3.9. For the sake of convenience, vertices are numbered. The heuristic function of vertex 1 is equal to $\hat{h}_1(v) = 7$, since two cubes are located in wrong places (2 and 3). Places that they should occupy are occupied by other cubes and two cubes are located on top of cube 2,

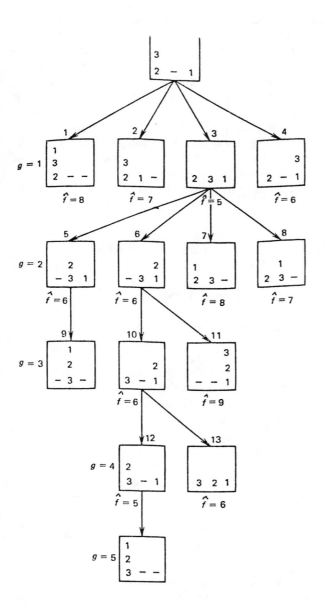

Figure 3.9 Solution graph of the scene-transformation problem.

and there is one cube on top of cube 3 $\hat{h}(1) = 2 + 2 + 2 + 1 = 7$. Therefore the estimating function $\hat{f}(1) = g(1) + \hat{h}(1) = 8$. In a similar manner, the values of the estimating function at other vertices of the first level can be computed.

Among the vertices of the first level, vertex 3 has the least value of function $\hat{f}$. Therefore, this vertex is uncovered in step 2. In so doing, repetitive vertices are not constructed and values of function $\hat{f}$ of such vertices are preserved, since the function $\hat{g}$ increases with every step, while function $\hat{h}$ does not vary. In step 3 a problem arises on how to select a vertex that should be uncovered during this step, since the estimating function $\hat{f}$ assumes equal minimal values at three vertices 5, 6, and 4. Let us assume that for equal values of $\hat{f}$, preference is given to vertices with a smaller $\hat{h}$. Vertices 5 and 6 have the same values of $\hat{f}$ and $\hat{h}$. The solution graph shown in Fig. 3.9 corresponds to a case when these vertices are arranged in such a manner that vertex 5 is uncovered first.

Let us examine certain properties of heuristic algorithms of an orderly sorting. An algorithm is called "guaranteeing" or "permissible" of for a random graph of states it accomplishes the construction of an optimum path.

It can be shown [1] that an orderly-sorting algorithm that uses an estimating function of the form 3.1 is a guaranteeing algorithm, if for all its vertices v the condition $\hat{h}(v) \leq h(v)$ is valid and the costs of all arcs exceed some positive number.

To compare the efficiency of sorting algorithms, a concept of "heuristic force" is used. Let A_1 and A_2 be random guaranteeing sorting algorithms and $f_1 = g(v) + \hat{h}_1(v)$ and $f_2 = g(v) + \hat{h}_2(v)$ are estimating functions used in them. The algorithm A_1 is called heuristically stronger than algorithm A_2 if at all vertices, except the target vertex, the inequality $\hat{h}_1(v) > \hat{h}_2(v)$ is satisfied. A heuristically stronger algorithm A_1 uncovers fewer vertices during a search for a minimum path than the A_2 algorithm. Let us denote by B a set of guaranteeing algorithms that may be used to solve a given problem. The algorithm $A^* \in B$ is called the "optimum" if it does not uncover more vertices than any other algorithm $A \in B$.

Optimization of the kind described above is not a conclusive indication of the degree of efficiency of an algorithm, since it does not take into account the complexity of computation of the heuristic function $\hat{h}(v)$. In most practical problems, a more objective characteristic is the estimate of the volume of computations required for determining values of $\hat{h}(v)$ at all steps on the way to the target vertex.

3.3   SOLUTION SEARCH WHEN PROBLEMS ARE REDUCED TO SUBPROBLEMS

The solution search for a given type of problem representation is based on a graph of problem reduction that is a graph of the type AND-OR. Implicitly, an AND-OR graph is defined by means of the description of operators of subproblem generation. An operator transforms the initial problem description into a set of filial subproblems. This transformation should be such that the solution of all filial subproblems ensures the solution of the initial problem. If the set of filial problems consists of one element, one problem is substituted by an equivalent problem. For a specific problem, there may be a set of operators of transformation. The application of each operator generates an alternate set of subproblems, which determines the existence of the relationship OR in the reduction graph.

The construction of a problem reduction graph is analogous to the construction of a solution search graph in the state space. The purpose of a search is to show that the initial problem is solvable. The search procedure may be viewed as the construction of a solution tree. The solution tree is a subtree (subgraph) of a problem reduction graph with a root in the initial vertex and consisting of solvable vertices.

A search in a problem reduction graph differs from that in a state graph in that it includes procedures for verifying the solvability or insolvability of vertices instead of verification procedures of the correspondence of a state to the target state. A vertex is insolvable if it is a dead-end vertex or if all its filial vertices are insolvable when they are linked by the OR relationship, or if at least one of its filial vertices is insolvable when they are linked by the AND relationship. The reduction process (vertex uncovering) ends when solvability or insolvability of the initial vertex is established.

As in the case of the state space search, there are methods of blind and orderly sorting in the reduction graph.

The complete-sorting algorithm. In the method of a complete sorting, vertices are uncovered in the order in which they are constructed. As was pointed out previously, a complete-sorting algorithm during a search of a solving graph has a specificity determined by verification procedures as to whether the vertices are solvable or insolvable, and is formulated as follows [1]:

1. Enter the initial vertex s_0 into the OPN list.

2. Take the first vertex from the OPN list and enter it into the CSD list; denote this vertex as v.

3. Uncover vertex v and enter all its filial vertices at the end of the OPN list and draw from them pointers to vertex v. If there are no filial vertices, then enter the vertex as an insolvable and go to the next step; otherwise, go to step **7**.

4. Apply the procedure of marking insolvable vertices to the search tree.

5. If the initial vertex is marked as insolvable, output a failure message; otherwise, go to the next step.

6. Delete from the OPN list all vertices that have insolvable vertices that precede them and go to step **2**.

7. If filial vertices are linked, then go to the next step; otherwise, go to step **9**.

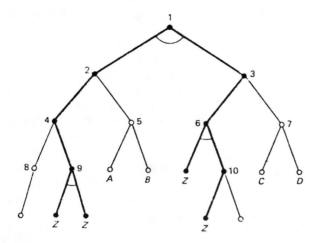

Figure 3.10   Reduction search tree using the complete-sorting algorithm.

8. If all filial vertices are terminal, then enter v as solvable and go to step **10**, otherwise, enter it as insolvable and go to step **4**.

9. If at least one filial vertex is terminal, then mark v as solvable and go to step **10**, otherwise, mark it as insolvable and go to step **4**.

10. Mark solvable vertices.

11. If the initial vertex is marked as solvable, then output the solution tree; otherwise, go to the next step.

12. Delete from the OPN list all vertices that are solvable or have solvable vertices that precede them, and go to step **2**.

Figure 3.10 depicts an example of a search tree for the reduction problem. Its vertices are marked with numbers indicating sequence of the vertex uncovering with the aid of a complete-sorting algorithm. Terminal vertices are denoted by Z. The solution tree is drawn in bold lines. Vertices A and B are not uncovered since they are dead-end vertices. Vertices C and D may be neither dead-end nor terminal. However, they are not uncovered since after uncovering vertex *10* the solution tree was established and the complete-sorting algorithm stopped the search.

The in-depth sorting algorithm.   Using the in-depth sorting algorithm, a solution tree is sought within the limits of a specified depth and during each step a vertex is uncovered whose depth is less than that of a vertex that was uncovered immediately before it. If the solution tree contains vertices that exceed the limit depth, the solution will obviously not be found.

The in-depth sorting algorithm includes the following step sequence [1]:

1. Enter the initial vertex s_0 into the OPN list.

2. Take the first vertex from the OPN list and enter it into the CSD list; mark this vertex by v.

3. If the depth of vertex v is equal to the limit depth, mark vertex v as insolvable and go to step **5**; otherwise, go to the next step.

4. Uncover vertex v. Enter all filial vertices (in an arbitrary order) at the beginning of the OPN list and draw pointers to vertex v from them. If there are no filial vertices, mark vertex v as insolvable and go to the next step; otherwise, go to step **8**.

5. Apply the insolvable vertices procedure to the search tree.

6. If the initial vertex is marked as insolvable, output a message that the sorting failed; otherwise, go to the next step.

7. Delete all vertices that have insolvable vertices that precede them from the OPN list. Go to step **2**.

8. If filial vertices are linked, go to the next step; if not, go to step **10**.

9. If all filial vertices are terminal, mark v as solvable and go to step **11**; otherwise, go to step **5**.

10. If at least one filial vertex is terminal, mark v as solvable and go to step **11**; otherwise, mark it as insolvable and go to step **5**.

11. Apply the marking procedure of solvable vertices to the sort tree.

12. If the initial vertex is marked as solvable, the solution tree is outputted; if not, go to the next step.

13. Delete all vertices that are solvable or have solvable vertices that precede them from the OPN list and go to step **2**.

Figure 3.11 shows the search tree of problem reduction that is constructed during an in-depth sorting with the limit depth equal to 4. The solving tree is drawn in bold lines. Numbers at vertices indicate the sequence in which the vertices are uncovered. Vertex F is not uncovered, although it is neither a dead-end nor terminal and has not reached the limit depth, since before its turn to be uncovered it was determined that its preceding vertex *2* is solvable and, in accordance with step **11**, it was deleted from the OPN list. Vertex G is not uncovered, since the solving tree has been found.

During a transition to a tree of the type AND-OR, a singularity arises when repeating vertices appear, i.e., vertices that are equivalent (they are determining

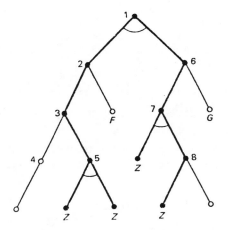

Figure 3.11   Reduction search tree using the in-depth sorting algorithm.

the same subproblem) to those that were constructed previously. In modified complete-sorting and in-depth sorting methods, repeating vertices are not constructed on a state tree. This cannot be done during sorting on a tree of the AND-OR type. Figure 3.12 shows an example of a search tree of the AND-OR type constructed during a complete sorting process. Vertex *6* in this tree is repeating; it is equivalent to vertex *3*. It is obvious that the repeating vertex *6* should not be constructed, since otherwise it would be impossible to obtain a solution tree. In this example it is not mandatory to uncover the repeating vertex *6* (if in the sorting algorithm there is a procedure for establishing the equivalency of various vertices), since at the moment of its uncovering it is known that the equivalent vertex *3* is solvable.

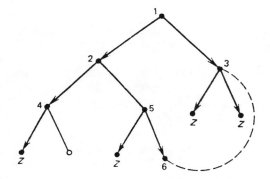

Figure 3.12   Problem reduction tree with a repeating vertex.

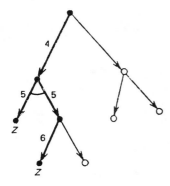

Figure 3.13   Solution tree for determining the cost of the route.

Heuristic search (sorting) methods. To orderly arrange uncovered vertices on a tree of the type AND-OR, the so-called optimum potential solution tree is used. A heuristic function is used to identify it. During a solution search in the state space, the heuristic function was defined as an estimate of cost of the optimum path from the given vertex to the target vertex. Cost estimate of the solution tree is used in trees of the AND-OR type.

Cost of the solution tree. There are two types of costs: a total cost that represents the sum of costs of all arcs in the solution tree, and a maximum cost that equals the cost of path between two vertices of the solution tree with the maximum cost. The cost of path is defined as the sum of arc costs making up this path. These definitions may be explained by using an example of the solution tree shown in Fig. 3.13 in bold lines, where arc numbers indicate their costs. In this example, the total cost is 20 and the maximum cost is 15.

If the cost of arcs equals unity, then total cost is equal to the number of arcs in the solution tree, and the maximum cost equals the number of arcs along the path between the two most remote vertices in the solution tree.

A solution tree that has a minimum cost (total or maximum, depending on which constitutes the optimization criterion) is called "optimum." Let $h(s_0)$ be the cost of an optimum solution tree with the root in the initial vertex s_0, and $h(v)$ be the minimum cost of a solution tree (or subtree of the solution tree) with vertex at point v. The cost $h(v)$ may be defined in the following manner (recursive definition):

1)   if v is a terminal vertex, then $h(v) = 0$;
2)   if v is a nonterminal vertex and has free filial vertices (vertices of the type OR) $v_1, \ldots, v_k$, then

$$h(v) = \min_{i \in (1,\ldots,k)} [c(v, v_i) + h(v_i)]$$

where $c(v, v_i)$ is the cost of the arc between vertices v and v_i; if v is a nonterminal vertex and has linked filial vertices $v_1, \ldots, v_k$, then total and maximum costs are defined in accordance with the following formulae

$$h(v) = \sum_{i=1}^{k}[c(v, v_i) + h(v_i)]$$

$$h(v) = \max_{i \in (1,\ldots,k)}[c(v, v_i) + h(v_i)]$$

For dead-end vertices v, function $h(v)$ is indeterminate.

Heuristic function. This is the name given to function $\hat{h}(v)$, which is the estimate of minimum cost $h(v)$. Let us see how a heuristic function may be constructed. A search tree constructed during a sorting process, during each step until the construction of the problem reduction tree is completed, has vertices that have no constructed filial vertices. Such vertices are called "end" vertices. They may be terminal, dead-end, and uncovered, i.e., vertices for which at the moment no filial vertices have been constructed. For end vertices v, the heuristic function is constructed as follows: if v is a terminal vertex, then $\hat{h}(v) = 0$; if v is an uncovered vertex, then $\hat{h}(v)$ is constructed as an estimate of function $h(v)$ on the basis of heuristic information associated with the initial problem.

For non-end vertices, a heuristic function is constructed as follows:

- if v has free filial vertices $v_1, \ldots, v_k$, then

$$\hat{h}(v) = \min_{i \in (1,\ldots,k)}[c(v, v_i) + \hat{h}(v_i)]$$

- if v has linked filial vertices $v_1, \ldots, v_k$, then heuristic functions for total and maximum costs are determined in accordance with the formulae

$$\hat{h}(v) = \sum_{i=1}^{k}[c(v, v_i) + \hat{h}(v_i)]$$

$$\hat{h}(v) = \max_{i \in (1,\ldots,k)}[c(v, v_i) + \hat{h}(v_i)]$$

For dead-end vertices, function $\hat{h}$ is not determinated.

Heuristic algorithm of orderly sorting. The search tree contains a set of sub-trees with the root at the initial point, each of which could be the initial part of a solution tree. These subtrees are called "potential" solution trees. A potential solution tree D_0 is called "optimum" if it has the following property: if vertex v, contained in tree D_0, has linked filial vertices in the sorting tree, then all of them are contained in the tree D_0; if this vertex has free filial vertices in the sorting tree

$v_i (i = 1, \ldots, k)$, then the tree D_0 contains one of them, which has the minimum value of $[c(v, v_i) + \hat{h}(v_i)]$.

In the heuristic method of orderly sorting it is assumed that the optimum potential solution tree is the initial part of the optimum solution tree and that its uncovered end vertices are being uncovered first. To derive the tree D_0 from the solution tree during a sorting procedure, it is necessary after each subsequent uncovering of a vertex to compute the heuristic function $\hat{h}$ for each of the newly constructed vertices and for each vertex preceding the just-uncovered vertex. For the remaining vertices, the value of $\hat{h}$ does not change.

The heuristic algorithm of orderly sorting may be formulated as follows:

1. Enter the initial vertex s_0 into the OPN list and compute $\hat{h}(s_0)$.

2. Determine the optimum potential solution tree D_0;

3. Choose some end vertex of the tree D_0 which is in the OPN list and enter it into the CSD list. Mark this vertex v.

4. If v is a terminal vertex, then enter it as solvable and go to the next step. Otherwise, go to step **8**.

5. Apply the marking procedure for solvable vertices to the tree D_0.

6. If the initial vertex is marked as solvable, output the tree D_0 as the solution tree; otherwise go to the next step.

7. Delete from the OPN list all vertices that have solvable vertices that precede them and go to step **2**.

8. Uncover vertex v. If it has no filial vertices, mark it as insolvable and go to the next step; otherwise, go to step **12**.

9. Apply the marking procedure of insolvable vertices to the tree D_0.

10. If the initial vertex is marked as insolvable, then issue a failure message; otherwise, go to the next step.

11. Delete from the OPN list all vertices that have insolvable vertices that precede them and go to step **2**.

12. Enter all filial vertices into the OPN list and draw from them pointers back to vertex v. Compute values $\hat{h}$ for these vertices. Recompute values $\hat{h}$ for vertex v and for vertices that precede it.

13. Go to step **2**.

In this algorithm of orderly sorting, it is assumed that a specific order of uncovering end vertices of tree D_0 contained in the OPN list is assigned. In establishing this order, the following considerations [1] may be used. End vertices of tree D_0 are uncovered first, since it is assumed that D_0 is the initial part of the optimum tree. In actuality, it may not be so. In such cases, it is desirable that the error in the selection of the tree D_0 is discovered as early as possible. In the case when the D_0 tree is indeed the initial part of the optimum tree, the order of uncovering its end vertices does not matter, since all of them must be uncovered. Therefore, that end vertex of the tree which will help discover the indicated error with the highest probability should be uncovered. In particular, if total cost is used as the optimization criterion, then such a vertex may be the end vertex of the tree from the OPN list with the largest value of $\hat{h}$.

Guaranteeing heuristic sorting algorithms. As in the case of sorting in the state space, a sorting algorithm for the tree AND-OR will be referred to as "guaranteeing" if its use always leads to the discovery of an optimum solution tree (provided that it exists).

It can be shown [1] that if for any vertex v from the OPN list, the heuristic function satisfies the inequality $\hat{h}(v) \leq h(v)$ and costs of all arcs are larger than some value $\delta > 0$, then the heuristic sorting algorithm corresponding to function $\hat{h}$ will be guaranteeing.

3.4   PROBLEM SOLUTION METHOD BASED ON PROOF OF THEOREMS

As was noted in §2.2, predicate calculus is widely used in artificial intelligence for solving problems. The problem solution method that utilizes techniques of predicate logic is based upon the problem representation in the form of a theorem: the formula $\mathcal{F}$ logically follows from a set of formulae $\Phi_0 (\Phi_0 \Rightarrow \mathcal{F})$. The proof of this theorem consists of showing that every interpretation satisfying Φ_0 also satisfies $\mathcal{F}$ or, which is the same, that the combination $\Phi_1 = \Phi_0 \cup \mathcal{F}$ cannot be satisfied. Usually the second approach is used, i.e., one that shows that set Φ_1 cannot be satisfied.

Identity transformations. During the theorem proof, all formulae in Φ_1 are presented in the form of disjunctions of literals. A literal (or a litera) is any elementary formula or its negation. A formula representing a disjunction of literals is called a "conjecture" or a "disjunct." In other words, in proving the theorem, formulae in Φ_1 are represented in the form of conjectures. In doing so, "identity transformations," that is, transitions from some formulae to their equivalents, are used.

It is obvious that all equalities that take place for the propositional forms (in the propositional calculus) and follow from the property of logical operation (including equalities following from the de Morgan theorem and theorems 2.2 and 2.3 concerning the representation of a random propositional form as normal forms) remain in force for the formulae (in the predicate logic) if these latter do not contain quantors. Let us point out the main properties of quantors, i.e., let us write equalities that directly follow from their definition. After replacing the predicates by the formulae from 2.12 and 2.13, we get

$$\overline{\forall x \mathcal{F}(x)} = \exists x \overline{\mathcal{F}}(x), \quad \overline{\exists x \mathcal{F}(x)} = \forall x \overline{\mathcal{F}}(x) \tag{3.2}$$

If the formula $\mathcal{F}$ does not contain an object variable x, then

$$\left. \begin{array}{l} \forall x(\mathcal{F} \wedge \mathcal{G}(x)) = \mathcal{F} \wedge \forall x \mathcal{G}(x) \\ \forall x(\mathcal{F} \vee \mathcal{G}(x)) = \mathcal{F} \vee \forall x \mathcal{G}(x) \\ \exists x(\mathcal{F} \wedge \mathcal{G}(x)) = \mathcal{F} \wedge \exists x \mathcal{G}(x) \\ \exists x(\mathcal{F} \vee \mathcal{G}(x)) = \mathcal{F} \vee \exists x \mathcal{G}(x) \end{array} \right\} \tag{3.3}$$

In addition, let us write a number of obvious equalities

$$\left. \begin{array}{r} \forall x(\mathcal{F}(x) \wedge \mathcal{G}(x)) = \forall x \mathcal{F}(x) \wedge \forall x \mathcal{G}(x) \\ \forall x \forall y \mathcal{F}(x, y) = \forall y \forall x \mathcal{F}(x, y) \end{array} \right\}$$

$$\exists x(\mathcal{F}(x) \vee \mathcal{G}(x)) = \exists x \mathcal{F}(x) \vee \exists x \mathcal{G}(x)$$

$$\exists x \exists y \mathcal{F}(x, y) = \exists y \exists x \mathcal{F}(x, y) \tag{3.4}$$

As follows from the definition, in a formula represented in the form of a conjecture, only disjunction and negation operations (signs) are explicitly used. In so doing, each negation sign is applied to no more than one predicate character. Therefore, for the representation of an arbitrary formula in Φ_1 in the form of conjecture, it is necessary to exclude all other logical operations (including quantors) and to reduce the operating interval of the negation sign to one predicate character.

Let us sequentially examine identity transformations that must be performed in the process of transforming a formula into conjectures [1].

1. *The exclusion of implication signs.* The implication sign may be excluded using the equality $(A \rightarrow B) = (\overline{A} \vee B)$. From this equality of propositional calculus, there follow, for instance, equalities

$$\exists x(F(x) \rightarrow \forall y G(y)) = \exists x(\overline{F}(x) \vee \forall y G(y))$$

$$\forall x F(x) \rightarrow [G(y) \rightarrow \forall x H(x)] = \forall x F(x) \vee [\overline{G}(y) \vee \forall x H(x)]$$

$$[\exists x F(x) \rightarrow \forall y G(y)] \rightarrow H(z) = \overline{[\exists x F(x) \rightarrow \forall y G(y)]} \vee H(z)$$

$$= \overline{[\overline{\exists x F(x)} \vee \forall y G(y)]} \vee H(z)$$

2. *Reduction of the operational interval of negation signs.* It is possible to reduce the operational interval of negation signs by utilizing the following equalities

$$\overline{F \wedge G} = \overline{F} \vee \overline{G}, \quad \overline{F \vee G} = \overline{F} \wedge \overline{G}, \quad \overline{\overline{F}} = F$$

$$\overline{\forall x F(x)} = \exists x \overline{F}(x), \quad \overline{\exists x F(x)} = \forall x \overline{F}(x)$$

Using these equalities, for instance, we obtain

$$\overline{\exists x F(x) \vee \forall y G(y)} = \overline{\overline{\exists x F(x)}} \wedge \overline{\forall y (G)(y)} = \exists x F(x) \wedge \exists y \overline{G}(y)$$

3. *Standardization of variables.* In the region of quantor operation, their associated variable can be replaced by a random variable that does not coincide with some other variable contained in the operational interval of these quantors. For instance

$$\forall x F(x, y) \vee \exists x G(x) = \forall z F(z, y) \vee \exists u G(u)$$

However, the formulae $\forall x F(x, y)$ and $\forall y F(y, y)$ are not equivalent. The renaming of the associated variables of the formula after which each quantor has its own associated variable (different from others) is called "standardization of variables."

4. *Exclusion of quantors of existence.* In the formula $\forall x (\exists y G(x, y))$, which can be interpreted, for instance, as "for all x there exists such y that x is no larger than y," quantor $\exists y$ is located within the region of operation of quantor $\forall x$. Therefore, y which "exists" may depend on x. Let us suppose that this dependence is determined in an explicit form with the aid of function $g(x)$, which transforms every value of x into y. Such a function is called the "Skolem function." For denoting the Skolem function, the functional characters that are already used inside the formula should not be used. If the quantor of existence is located within the region of operation of two or more quantors of commonality, then the Skolem function will respectively depend on two or more arguments.

If the existence quantor being excluded belongs to no region of operations of a commonality quantor, then the Skolem function does not contain an argument, i.e., it is a constant. Thus, the formula $\exists x F(x)$ after the exclusion of the existence quantor is transformed into the formula $F(a)$, where a is a constant which we know "exists."

5. *The exclusion of commonality quantors.* After the exclusion of existence quantors and standardization of variables, the formula contains only commonality quantors, each with its own variable. Therefore these quantors

may be transferred to the first part of the formula and the part of the formula located behind each quantor may be viewed as the region of its operation. For instance

$$\forall x \{F(x) \wedge \forall y[G(y) \vee H(x,y)]\}$$

$$= \forall x \forall y \{F(x) \wedge [G(y) \vee H(x,y)]\}$$

Due to the fact that in theorem $\{\Phi_0 \Rightarrow \mathcal{F}\}$ the set consists of closed formulae, i.e., formulae that do not contain free variables, and that the formula $\mathcal{F}$ is closed, all variables in the formula $\Phi_1 = \Phi_0 \cup \overline{\mathcal{F}}$ belong to commonality quantors. And since the order in which commonality quantors are arranged does not matter, these quantors may not be explicitly indicated, i.e., they may be excluded, having agreed that all variables in the formulae belong to commonality quantors.

6. *Formula representation in a completed conjunctive normal form (CCNF).* After excluding all quantors, (per theorem 2.3), the formula (from Φ_1) may be transformed into a CCNF, i.e., presented as a conjunction of a finite set of conjunctive terms K_i. By definition, a conjunction term represents disjunction of literals [predicates and/or their negations] and is a conjecture. Consequently, a formula in CCNF represent conjunctions of a finite set of conjectures. Since some interpretation would satisfy a formula of the form $(K_1 \wedge K_2 \wedge \ldots \wedge K_n)$ in that (and only that) case when it satisfies formulae K_1, K_2, $\ldots$, K_n simultaneously, then the initial formula in Φ_1 may be replaced by set $\{K_i, i = 1, \ldots, n\}$ of conjunctive terms (conjectures).

Let us consider an example of transforming a formula into conjectures [1]. Suppose we are given a formula

$$\forall x \{P(x) \rightarrow \{[\forall y[P(y) \rightarrow P(f(x,y))]]$$

$$\wedge [\overline{\forall y[Q(x,y) \rightarrow P(y)]}]\}\}$$

Let us exclude the implication signs

$$\forall x \{\overline{P}(x) \vee \{\forall y[\overline{P}(y) \vee P(f(x,y))]]$$

$$\wedge [\overline{\forall y[\overline{Q}(x,y) \vee P(y)]}]\}\}$$

Let us reduce operational intervals of negation sign to one predicate

$$\forall x \{\overline{P}(x) \vee\} \{\forall y[\overline{P}(y) \vee P(f(x,y))]]$$

$$\wedge [\exists y[Q(x,y) \wedge \overline{P}(y)]]\}\}$$

Let us standardize variables

$$\forall x \left\{ \overline{P}(x) \vee \left\{ \forall y [\overline{P}(y) \vee P(f(x,y))]] \right. \right.$$

$$\wedge [\exists z [Q(x,z) \wedge \overline{P}(z)]]\} \}$$

Let us exclude all quantors

$$\overline{P}(x) \vee \left\{ [P(y) \vee P(f(x,y))] \right.$$

$$\wedge [Q(x,g(x)) \wedge \overline{P}(g(x))]\}$$

Here $g(x)$ is the Skolem function.

The existence quantor $\exists z$ is located in the region of operation of only one commonality quantor $\forall x$, so the Skolem function depends only on x. Using the distributive property (law) of disjunction with respect to a conjunction, the last formula is easily reduced to a CCNF, from which we obtain the following conjectures

$$\overline{P}(x) \vee \overline{P}(y) \vee P(f(x,y)); \qquad \overline{P}(x) \vee Q(x,g(x));$$

$$\overline{P}(x) \vee \overline{P}(g(x))$$

The Erbran universe. Let us suppose that the process of transforming of formulae from Φ_1 into conjectures is completed and we obtained some set Φ consisting only of conjectures, i.e., formulae representing disjunctions of literals. To prove that set Φ cannot be completed means that it is shown that there is no interpretation that can satisfy it. To specify an interpretation, as was pointed out previously, means the following:

1)   the assignment of the interpretation region **M**;
2)   the determination of connections (correspondence) between: a) an object constant in Φ and a separate element from **M**; b) a functional character in Φ and a function in **M**; c) a predicate character in Φ and a specific relationship among elements from **M**.

It is obvious that it is impossible to list all possible regions of interpretation and connections: first, there may be an infinite number of them and, second, it is not clear how they should be constructed. Nevertheless, the problem of proving the theorem can be solved. The solution method given below is based on the following statement [1]: if set Φ cannot be completed in the region $H(\Phi)$ which is called the "Erbran universe," then it is incompletable in any region.

Let us define the region $H(\Phi)$ and explain the meaning of saying that set $H(\Phi)$ is incompletable in this region. The Erbran universe $H(\Phi)$ for a set of conjectures Φ is defined as follows (a recursive definition):

1. The set of all object constants mentioned in Φ belongs to the region $H(\Phi)$. If Φ does not contain object constants, then an arbitrary object constant is included into $H(\Phi)$, for instance, a.

2. If therms $t_1, \ldots, t_n$ belong to $H(\Phi)$ and f_i is any n-place functional character mentioned in Φ, then regions $H(\Phi)$ also belong to $f_i(t_1, \ldots, t_n)$.

3. No other therms belong to the $H(\Phi)$ region.

As follows from the definition, the Erbran universe $H(\Phi)$ is a set of constant therms, i.e., therms not containing object variables. In general, set $H(\Phi)$ is infinite but denumberable since its elements may be numbered.

Suppose, for instance, that $\Phi = \{ P(x) \vee Q(a) \vee \overline{P}(f(x)), \overline{Q}(b) \vee P(g(x,y)) \}$. Set Φ contains object constants a and b and functional characters f and g. Region $H(\Phi)$ is a denumerable set $\{ a, b, f(a), f(b), g(a,a), g(a,b), g(b,a), g(b,b), f(f(a)), f(f(b)), g(a,f(a)), \ldots \}$.

The expression which is obtained by substituting constant therms into an arbitrary formula is called its "constant special case." If in the region $H(\Phi)$ an interpretation is given for conjectures in the set Φ, truth values T or F are assigned to constant special cases of elementary formulae in Φ, i.e., the identification of elementary formulae in Φ is performed. The identification of an elementary formula $F(x_1, \ldots, x_n)$ is performed by assigning values of T or F independently to all constant special cases which are obtained as a result of the substitution of elements from $H(\Phi)$ instead of variables $x_1, \ldots, x_n$. The set of constant special cases for all elementary formulae in Φ obtained as a result of the substitution of elements from $H(\Phi)$ instead of variables in Φ is called the "Erbran base" for Φ. Elements from this set are called "atoms." Specification of interpretation in $H(\Phi)$ of conjectures in Φ corresponds to assigning values T and F to all atoms. The set is incompletable in $H(\Phi)$ if any interpretation in $H(\Phi)$ does not satisfy Φ.

Semantic tree. In general, the Erbran base is an infinite but denumerable set, so that its elements may be numbered and an Erbran base may be represented as a sequence $(p_1, p_2, \ldots)$.

Specifying interpretations in $H(\Phi)$ is conveniently done using the so-called semantic tree. Let us consider a "binary tree" (e.g., a tree each of whose vertices has exactly two filial vertices). It is constructed as follows. From the initial vertex (the tree root), two arcs are drawn: left arcs correspond to the assignment to atom p_{i+1} of the truth value T (these arcs are marked by p_{i+1}); right arcs, of the truth value F (these arcs are marked by $\overline{p}_{i+1}$). This process is continued until each element of the Erbran base has a truth value.

A tree constructed in this manner is called a "semantic tree." Each path in this tree from root to a terminal vertex gives one of the interpretations in $H(\Phi)$ for set Φ. Arc markings of this path determine truth values assumed by atoms of the Erbran base for a given interpretation. A set of paths from the root of a semantic tree to all its terminal vertices determines all possible interpretations that may be specified in the Erbran universe for set Φ. If we write arc markings of some path from the root of the semantic tree to its terminal vertex, we will obtain a sequence of the form $(g_1, g_2, \ldots)$, where $g_i (i = 1, 2, \ldots)$ denotes p_i or $\overline{p}_i$. This sequence is called a "model" for a given set of conjectures Φ. Each model corresponds to a specific interpretation. A model does not satisfy a conjecture if there exists a constant special case of this conjecture that has a value F for identification determined by this model.

As an example, let us consider a set [1]

$$\Phi = \left\{ P(x) \vee Q(y), \quad \overline{P}(Q), \quad \overline{Q}(b) \right\} \tag{3.5}$$

Finite set $H(\Phi) = \{a, b\}$ is the Erbran universe. The Erbran base is also finite. Let us arrange it as follows

$$\left\{ P(a), \quad Q(a), \quad P(b), \quad Q(b) \right\}$$

A semantic tree corresponding to this set is shown in Fig. 3.14. Let us denote through M_i a model corresponding to a path from the root to the i-th terminal vertex. The model $M_i = \left\{ P(a), Q(a), P(b), Q(b) \right\}$ satisfies none of conjectures $\overline{P}(a)$ and $\overline{Q}(b)$. Model $M_4 = \left\{ P(a), Q(a), \overline{P}(b), \overline{Q}(b) \right\}$ does not satisfy the conjecture $P(x) \vee Q(y)$, since the constant special case $P(b) \vee Q(b)$ in this model has the value F. A model does not satisfy Φ when and only when Φ contains at least one conjecture that is not satisfied by this model. Therefore, the proof of incompletability reduces to a proof that for any model determined by the semantic tree there exists a conjecture in Φ that this model does not satisfy. By examining all models determined by a semantic tree shown in Fig. 3.14, it is not difficult to show that set 3.5 is incompletable.

Closure of a semantic tree for an incompletable set of conjectures. To ensure that some interpretation does not satisfy a set of conjectures, a path to the terminal vertex does not have to be traced. In the example shown in Fig. 3.14, by the time one arrives at the left vertex of the first level, it may be seen that none of the eight possible interpretations in which $P(a)$ has the value T satisfies the set 3.5, since they do not satisfy conjecture $\overline{P}(a)$ from this set. Therefore, there is no need to trace the left part of the tree. In Fig. 3.14, vertices which permit establishing for the first time that some interpretations do not satisfy Φ are marked by darkened circles. Such vertices are referred to as "adverse" or "adverse for corresponding conjectures." Paths break off at the adverse vertices. A semantic tree for a set of conjectures Φ, all of whose paths break off at the adverse vertices, is called "closed" for Φ.

If a set of conjectures cannot be satisfied, then even under the condition that its Erbran base is infinite, each possible interpretation must break off at some adverse vertex. Indeed, if some path does not break off at an adverse vertex, then one can travel along this path for an infinitely long time. This would mean that the interpretation corresponding to this path satisfies the set of conjectures, which contradicts to the incompletability of this set. Thus, the semantic tree for an incompletable set of conjectures Φ is closed and contains a finite number of vertices located above the adverse vertices.

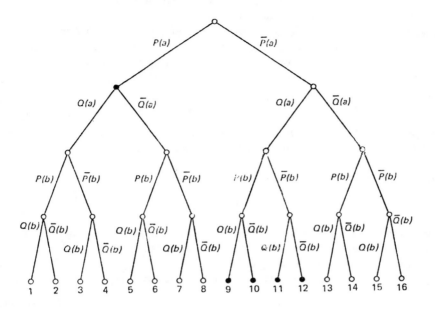

Figure 3.14 A semantic tree.

The resolution principle. In practice, the incompletability of a set of conjectures by constructing a semantic tree can seldom be established. Usually meaningful problems of proof are solved by means of the principle of resolution (or resolving).

Suppose there are conjectures of the form $P(x) \vee Q(y)$ and $R(z) \vee \overline{Q}(y)$. From these conjectures, conjecture $P(x) \vee R(z)$, which logically follows from the initial conjectures, can be derived

$$\{P(x) \vee Q(y), \quad R(z) \vee \overline{Q}(y)\} \Rightarrow \{P(x) \vee R(z)\} \tag{3.6}$$

The newly derived conjecture is called "resolvent." In general, it is possible to obtain several resolvents from one pair of conjectures. The process of obtaining resolvents is called "resolution."

Let us prove that if some interpretation satisfies initial conjectures, then it satisfies the resolution. Suppose some interpretation I satisfies the initial conjectures. Only two alternatives are possible:

1) the interpretation I satisfies $Q(y)$;
2) the interpretation I satisfies $\overline{Q}(y)$.

In the first case, the interpretation I will satisfy initial conjectures if it satisfies $R(z)$. Consequently, in this case, I satisfies the resolvent. In the second case, the interpretation I will satisfy initial conjectures if it satisfies $P(x)$. Consequently, I satisfies the resolvent in the second case as well.

If any interpretation satisfying all formulae (conjectures) of set Φ satisfies the formula (conjecture) $\mathcal{F}$, i.e., $\mathcal{F}$ logically follows from Φ, then $\mathcal{F}$ is called a "corollary" of Φ. A combination of rules used in obtaining corollaries — resolvents and factors (this concept is defined later) — from given conjectures is called the "principle of resolution."

The resolution principle includes the following two principles [3]:

1. The principle of syllogism in the propositional calculus consisting in the fact that from $(A \vee B)$ and $(\overline{A} \vee C)$, it logically follows that $(B \wedge C)$

$$[(A \vee B) \wedge (\overline{A} \vee C)] \Rightarrow (B \vee C)$$

   i.e., the propositional form $\{[(A \vee B) \wedge (\overline{A} \vee C)] \to (B \vee C)\}$ is a tautology.

2. The principle of the discovery of special cases in predicate calculus consisting of the fact that formula $\mathcal{F}(t_1, \ldots, t_n)$ obtained from $\mathcal{F}(x_1, \ldots, x_n)$ after substituting instead of x_i of arbitrary therms t_i is a special case of $\mathcal{F}(x_1, \ldots, x_n)$ and, consequently, of $\mathcal{F}(x_1, \ldots, x_n) \Rightarrow \mathcal{F}(t_1, \ldots, t_n)$.

Let us recall that conjectures are disjunctions of literals. Let us call literal L_1 "complementary" to literal L_2 if L_1 is a negation of $L_2 (L_1 = \overline{L_2})$. A constructive definition which can be used to obtain the resolvent can be given.

A resolvent of two conjectures is defined as their corollary obtained in the following manner:

1) variables of one conjecture are renamed so that they differ from variables of the other conjecture;

2) such a substitution is found for which some literal of one conjecture becomes complementary to some literal of the other conjecture, and this substitution is performed on both conjectures;

3) literals complementary to each other are deleted;

4) if there are identical literals, then all of them, except one in some conjecture, are deleted;

5) disjunctions of literals remaining in both conjectures is the resolvent.

A "factor" of some conjecture is the corollary of this conjecture obtained as follows. A substitution is determined for which some literals are identical; after completing this substitution, all literals except one are deleted; the disjunction of the remaining literals is the factor. The process of determining a factor is called "factorization."

Substitution. During resolution and factorization, a substitution that produces their special cases from some given formulae are employed. In general, any substitution used in the application of the resolution principle may be represented as a set of orderly arranged pairs $\theta = \{(t_i, x_i), \ldots, (t_n, x_n)\}$, where each pair (t_i, x_i) means that anywhere this substitution is made, variable x_i is replaced by the therm t_i. Applying, for instance, to literal $P(x, f(y), b)$ the substitution $\theta_1 = \{(z, x), (\omega, y)\}$, $\theta_2 = \{(a, y)\}$ and $\theta_3 = \{(g(z), x), (a, y)\}$, we obtain the corresponding special cases of the initial literal

$$P_{\theta_1} = P(z, f(\omega), b); \quad P_{\theta_2} = P(x, f(a), b)$$

$$P_{\theta_3} = P(g(z), f(a), b)$$

Here P_{θ_i} denotes a special case P obtained during the substitution of θ_i.

The "composition" of two substitutions of α and β is a formulation $\alpha\beta$ obtained during the application of the substitution of β to therms of the substitution α with the subsequent addition of any pairs from β containing variables which are not among the variables from α. For instance, if

$$\alpha = \{(f(x), z\} \quad \text{and} \quad \beta = \{(a, x), (b, y), (d, z)\}$$

then the composition

$$\alpha\beta = \{(f(a), z), (b, y)\}$$

It is not difficult to see that the application to the literal of a sequence of substitutions α and β produces the same result as the application of substitution $\alpha\beta : (P_\alpha)_\beta = P_{\alpha\beta}$. A composition of a formulation is associative

$$(\alpha\beta)\gamma = \alpha(\beta\gamma)$$

Unification. A set of literals $\{L_i\}$ is called "unifyable" if there exists such a substitution θ that $L_{1\theta} = L_{2\theta} = \dots$. The substitution θ in this case is called the "unificator" for $\{L_i\}$. A unificator λ for $\{L_i\}$ is called the "simplest" (or the "most general") if, regardless of the nature of the unificator θ for $\{L_i\}$, there will be found such a substitution δ that $\{l_i\}_{\lambda\delta} = \{L_i\}_\theta$. There exists an algorithm, called the unification algorithm, that permits the simplest unificator for the unifyable set of literals $\{L_i\}$ to be obtained.

The unification algorithm is recurrent. Its first substitution is empty (i.e., no substitution is performed in step **1**). Suppose that at the k-th step we obtain the substitution λ_k. If, during its application to the set $\{L_i\}$ all its literals become identical, then the simplest unificator $\lambda = \lambda_k$. Otherwise a substitution λ_{k+1} is determined as follows. By examining each of literals in $\{L_i\}_{\lambda_k}$ ($\{L_i\}_{\lambda_k}$ is a set of literals obtained from $\{L_i\}$ by substituting λ_k) as a chain of symbols (predicate and functional characters, constants, and variables), we identify the position of the first symbol in which not all literals have the same symbol. Next a set of disconcordance is constructed containing tamely constructed expressions of all literals (expressions) which begin at this position (a tamely constructed expression is either a therm or a literal). Thus, a set of disconcordance for $f\{P(a, f(b, x), g(y)), P(a, f(b, h(z)), u)\}$ is $\{x, h(z)\}$. Next, we determine the substitution λ_{k+1} as, for instance

$$\lambda_{k+1} = \lambda_k\{(t_k, x_k)\}$$

where x_k is some variable from the disconcordance set; t_k is a therm (possibly another variable) from the same set not containing x_k. If the disconcordance set does not contain variables at all, then set $\{L_i\}$ cannot be unified. For example, if not all predicate characters (i.e., the symbols in the first position) are identical, then the disconcordance set coincides with set $\{L_i\}$ and none of its elements will be a variable. Let us note that if λ_k is an empty substitution, then composition $\lambda_k\{(t_k, x_k)\}$ coincides with the substitution $\{(t_k, x_k)\}$.

It can be shown [1] that the described algorithm determines the simplest unificator if set $\{L_i\}$ is unifyable and if it gives a message of failure otherwise.

Proof of the theorem with the aid of the resolution principle. As was shown previously, the problem of proving the theorem $(\Phi_0 \Rightarrow \mathcal{F})$ is equivalent to the problem of proving the incompletability of the combination $\{\Phi_0 \cup \mathcal{F}\}$. The proof of the incompletability of random set of Φ of conjectures with the aid of resolution (or the resolution principle) is based on the following.

Suppose $\mathcal{R}(\Phi)$ is the combination of set Φ with the set of resolvents of all pairs of conjectures from Φ, $\mathcal{R}^2(\Phi) = \mathcal{R}(\mathcal{R}(\Phi))$, etc., and $\mathcal{R}^0(\Phi) = \Phi$. It is not difficult to see that if set Φ is incompletable, then $\mathcal{R}^i(\Phi)$ for any $i \geq 1$ is also incompletable and vice versa, if $\mathcal{R}^i(\Phi)$ for any $i \geq 1$ is incompletable, then $\mathcal{R}^k(\Phi)$ for any $k \geq 0$ (i.e., including Φ) is incompletable.

If set Φ is incompletable, then there exists such a finite number n that in $\mathcal{R}^n(\Phi)$ there is an empty conjecture. This latter is possible if in $\mathcal{R}^{(n-1)}(\Phi)$. There are two such conjectures K_1 and K_2 which, after applying a corresponding substitution to them, become complementary $(K_1 = \overline{K}_2)$ a resolvent of only such conjecture pairs is an empty conjecture. Consequently, $\mathcal{R}^{(n-1)}(\Phi)$ and, respectively, all $\mathcal{R}^i(\Phi)$, including Φ, are incompletable. Thus, in order to prove the incompletability of set Φ, it is sufficient to show that during the application of the resolution principle to set Φ and its corollaries, one obtains an empty conjecture.

The proof with the aid of resolution can be demonstrated using a graph called a "graph of rebuttal (disproof)." In general, the term "rebuttal" designates a proof of incompletability of an arbitrary set of conjectures.

To construct a graph of rebuttal, representing a tree, conjectures from Φ are recorded at the end vertices of this graph. If two conjectures located at some vertices are solvable (e.g., they have a resolvent), then their resolvent is written at a vertex which directly follows these vertices and which is connected to them by edges. A root of the disproof tree is an empty conjecture denoted by the English work "nil" (nil meaning nothing, zero). A disproof tree, unlike the previously considered trees, is constructed beginning with the end vertices, rather than from a root. For example, Fig. 3.15 shows a disproof graph of an incompletable set of conjectures

$$\{P(x) \vee Q(x), \overline{Q}(f(z)), \overline{P}(f(z)) \vee R(Z), \overline{R}(\omega)\}$$

A proof with the aid of a resolution is a special case with the application of the principle of resolution. By definition, the resolution principle is a rule of

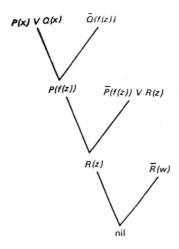

Figure 3.15 A disproof tree.

derivation of resolvents and factors. In general, the proof with the use of the resolution principle assumes the utilization of not only resolutions but also of factors, i.e., it includes resolution and factorization.

From the above, it follows that:

1. The resolution principle is complete in the sense of the following theorem that was proven for the first time by Robinson: if a finite set of conjectures is incompletable, then a disproof may be obtained by applying the resolution principle a finite number of times.

2. The resolution principle is noncontradicting in the sense that if, as a result of the application of the resolution principle, we obtain an empty conjecture, then the initial set of conjectures is mandatorily incompletable. In other words, the proof of incompletability of a set of conjectures with the aid of the resolution principle is correct (tamely).

Sorting strategies. A direct application of the resolution principle during which one sequentially constructs sets $\mathcal{R}(\Phi), \mathcal{R}^2(\Phi)$, etc. corresponds to a complete sorting. A complete sorting, as a rule, is impractical since sets $\mathcal{R}(\Phi), \mathcal{R}^2(\Phi)$ grow too rapidly. Therefore, in a practical proof one uses sorting methods or strategies that differ from the complete-sorting strategy by the use of different procedures accelerating sorting procedures. Such strategies include the so-called strategies of simplification, refinement, and arranging.

The simplification strategy. Sometimes a set of conjectures may be simplified by excluding certain conjectures from it or excluding certain literals from its conjectures. Such simplifications permit the growth speed of new conjectures to be reduced. Naturally, simplifications must be such that the simplified set would be incompletable when and only when the initial set is incompletable.

Any conjecture containing a literal and its negation is a tautology and it may be excluded, since any incompletable set containing a tautology remains incompletable after its removal.

A conjecture may be excluded if it contains a unique literal, i.e., a literal not contained with a negation in any conjecture. This is so because no resolution using such a conjecture or any conjecture derived from it would not lead to an empty conjecture. For instance, from a set of conjectures $\{P(x) \vee \overline{P}(b) \vee Q(y), \overline{P}(c) \vee P(y), \overline{P}(b) \vee P(x)\}$ one may exclude the first conjecture since it contains a unique literal $Q(y)$.

A conjecture $\{L_i\}$ is called a subcase of a set $\{M_i\}$ if there exists such a substitution θ that $\{L_i\}_0 \subseteq \{M_i\}$. For instance, $P(x)$ is a subcase of conjectures $P(y) \vee Q(z)$ and $P(a)$; $P(x) \vee Q(y)$ is a subcase of $P(f(a)) \vee Q(a) \vee R(z)$. From two conjectures of set Φ only the subcase may be retained with the condition that one conjecture is a subcase of the other one (without violating the incompletability properties).

Thus, simplification strategies include: a) strategy of excluding tautologies; b) strategy of excluding conjectures containing unique literal; c) strategy of employing subcases.

The refinement strategies. These strategies include methods of proof (disproof) that use not all but specific resolutions. Various refinement strategies differ from one another by the rule of selecting conjectures for resolutions.

In the majority of problems of theorem proofs, the incompletable set Φ may be broken down into two subsets: subset Φ_a consisting of axioms and subset Φ_0 consisting of negations of those conjectures that have to be proven. It is natural to assume that axioms are inherently noncontradicting, i.e., subset Φ_a is completable.

A "support" (or "carrying") set is a set of conjectures that includes: 1) all conjectures from subset Φ_0; 2) resolvents of those conjecture pairs, at least one of whom belongs to the support set.

A proof method which does not permit resolutions of two axioms (i.e., of two conjectures from Φ_0) or, which is the same, only permits those resolutions in which at least one conjecture from the support set participates, is called the "strategy of the support set." This is a complete strategy.

One other strategy that falls into the class of refinement strategies is the strategy of linear derivation. To define it, we introduce the concept of a least incompletable set that is closely related to the concept of the support set. A subset Φ^* of an incompletable set Φ is called the "least incompletable set" if it is incompletable while any of its own subsets is completable.

In general, the least incompletable set is not unique. For instance, if

$$\Phi = \{ P(x) \vee Q(x) \vee R(x), \ \overline{P}(x) \vee \overline{Q}(x), \ \overline{P}(x) \vee \overline{R}(x)$$

$$\overline{Q}(x) \vee \overline{R}(x), \ P(x), \ Q(x), \ R(x)\}$$

then for it the following least incompletable sets are possible

$$\Phi_1^* = \left\{ \overline{P}(x) \vee \overline{Q}(x), \ P(x), \ Q(x) \right\}$$

$$\Phi_2^* = \left\{ \overline{P}(x) \vee \overline{R}(x), \ P(x), \ R(x) \right\}$$

$$\Phi_3^* = \left\{ \overline{Q}(x) \vee \overline{R}(x), \ Q(x), \ R(x) \right\}$$

It should be noted that not all possible least incompletable sets are written here.

Let us number all resolvents in the order of their derivation $\mathcal{G}^i (i = 0, 1, 2, \ldots)$. The derivation of resolvents for which $\mathcal{G}^0$ is a resolvent that can be derived from some two conjectures belonging to a least incompletable set, while $\mathcal{G}^i (i \geq 1)$ is a resolvent that can be derived from two conjectures, one of which is resolvent $\mathcal{G}^{i-1}$, is called a "linear derivation." The proof method that uses the linear derivation is called the "strategy of linear derivation." This strategy is complete [4] in the sense that if Φ^* is the least incompletable set, then there exists at least one linear derivation of a nonempty conjecture (i.e., a linear derivation proving the incompletability of the set Φ^*) that utilizes conjectures from Φ^*. But one has to keep in mind that not all linear derivations that begin with a given conjecture from set Φ result in a nonempty conjecture.

The number of resolutions required for determining a disproof may be limited by using a model. This is the basis of a modeling strategy. Let us recall that a model is a specific interpretation in the Erbran universe, i.e., a combination of atoms of the Erbran base, each of which has a specific truth value.

A "model strategy" is a strategy where resolutions are applied only to such conjecture pairs among which at least one is not satisfied by the model. This strategy is based upon the following theorem.

Theorem 3.1 [1]. Suppose that Φ is a nonsatisfiable set of conjectures and M_1 is a model given on its Erbran base. Then there exists such a graph of disproof for Φ for which each of its vertices is either a conjecture from Φ, or has (as one of the directly preceding it vertices) a conjecture that is not satisfied by model M_1.

To what degree the number of required resolutions is reduced by using the model strategy depends on the chosen model M_1. The worse model M_1 is one which does not satisfy any conjecture from Φ, since the use of the model strategy with such a model does not decrease the number of required resolutions.

Closely related to the model strategy is a strategy which is called P_1-disproof. In any incompletable set Φ there is at least one conjecture without negations, i.e., a conjecture whose literals have no minus signs. Indeed, otherwise the model determined by a set of negations of atoms from the Erbran base would satisfy Φ, which would contradict the incompletability of Φ. A disproof for which each resolution is realized between two such conjectures, one of which is a conjecture without negations, is called the P_{1i}-disproof [1]. The completeness of this strategy follows from theorem 3.1. Indeed, a set of literals from Φ can be used as a model.

Here a conjecture within this model assumes a truth value when and only when it does not contain literals without negations.

Simplification strategies may also be considered as refinement strategies, since the application of subcases, exclusion of tautologies, etc. means that not all resolutions are used in simplification strategies, but there is a specific selection of conjectures for resolutions which is characteristic for refinement strategies.

Strategies of orderly arrangement. These include strategies in which no type of resolutions is forbidden. Rather, which one must be performed first is indicated.

Since the purpose of resolutions in the process of disproof is the derivation of an empty conjecture, then the strategy of preference for monomials suggests itself. In so doing, first the construction of a resolution for two monomials, i.e., conjectures containing only one literal, is attempted. If this succeeds, then an empty conjecture is obtained right away, i.e., a disproof. If there are no such monomials, then an attempt is made to find a resolvent for a pair monomial-binomial, then a pair monomial-trinomial, etc. As soon as some pair is solved, then the obtained resolvent is immediately compared with monomials in order to try to determine a resolvent for a pair monomial–the obtained resolvent.

Closely related to the strategy of monomial preference is the strategy of the "least number of components" that arranges resolutions in accordance with the number of literals in obtained resolvents. In this strategy, two conjectures producing a resolvent with the least number of literals are solved first.

Combined strategies. Naturally, it is not mandatory that the described strategies are used in their "pure" form, i.e., each strategy separately. It makes more sense to combine several strategies. In so doing, however, it must be kept in mind that although pure strategies are complete, combined strategies obtained on their basis may be incomplete [1].

In choosing a strategy, either pure or combined, expenses for additional calculations required for selecting corresponding resolutions may not be justified by economies obtained as a result of a decreased number of resolutions performed in the process of disproof.

Application of methods of a theorem proof to a problem solution. Many problems encountered in applications are in some cases reduced to a clarification whether some formula $\mathcal{F}$ logically follows from set Φ of formulae, while in other cases they are reduced to the establishment of value of element x (if it exists) for which a given formula $\mathcal{F}$, which contains x as one of its arguments, logically follows from Φ_0. In the second case it is first established whether formula $(\exists x \mathcal{F}(x))$ follows from Φ_0, and then, if the answer is affirmative, the corresponding value of x is determined.

Suppose the loading robot (lr) is located on the automated vehicle (av). The vehicle is located at the warehouse (ware). The question to be answered is where is the robot?

Let us present the problem as a problem of a theorem proof. To this end, let us introduce the following notation: P (predicate), to be in a specific location; $P(z,x)$, z is located at x. Using this notation, the problem statement (the factual part) may be written as a set of formulae

$$\Phi_0 = \{\forall x[P(\text{av}, x) \to P(\text{lr}, x)], \; P(\text{av}, \text{ware})\}$$

In set Φ_0, the first formula expresses the fact that regardless of where the vehicle is located, the loading robot is located at the same place; the second formula expresses the fact that the vehicle is located at the warehouse. The initial problem reduces to the proof of the theorem: $\Phi_0 \Rightarrow \{\exists x P(\text{lr}, x)\}$. After determining the corresponding value of x, we will obtain the answer to the question posed by the problem.

During the transformation of a question into the formula $\mathcal{F}$ that contains the existence quantor, it should be understood that the answer to the question will be a variable associated with this quantor. If on the basis of the given facts which can be presented as a set Φ_0, the answer to the posed question can be obtained, then the formula constructed in this manner will logically follow from Φ_0. After determining the theorem proof $\{\Phi_0 \Rightarrow \mathcal{F}\}$, the special case of a variable associated with the existence quantor that is the answer can be derived.

The proof that formula $\mathcal{F}$ follows from set Φ_0 is obtained in a normal way: formula $\overline{\mathcal{F}}$ is added to set Φ_0 and all terms of the expanded set $\Phi_1 = \Phi_0 \cup \overline{\mathcal{F}}$ are transformed into the conjectures format. Then, using the resolution principle, it is shown that set Φ_1 is incompletable.

Formula $\mathcal{F}$ which has to be proven (that it logically follows from Φ_0) is called a "supposition" and suppositions that are obtained from formulae contained in Φ_0 are called "axioms."

Let us construct a disproof tree for the considered example. Since $\overline{\exists x P(\text{lr}, x)} = \forall x P(\text{lr}, x) = P(\text{lr}, x)$, then the expanded set after transforming its formulae into conjectures has the form

$$\Phi = \{P(\text{av}, x) \vee P(\text{lr}, x), \; P(\text{av}, \text{ware}), \; P(\text{lr}, x)\}$$

The disproof tree is shown in Fig. 3.16.

The process of answer extraction. After constructing the disproof tree for extracting the answer to the posed question, a modified tree of a proof is constructed in the following manner:

1) to each conjecture following from a conjecture negation, its negation is added (in the sense of logical addition);
2) the same resolutions as during the construction of a disproof tree are performed.

After the completion of a modified tree proof, a special case of a conjecture that is used as the answer conjecture is obtained. A modified proof tree for this

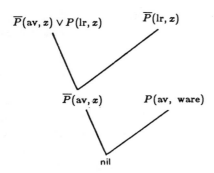

Figure 3.16 The disproof tree in the loading robot problem.

example is shown in Fig. 3.17. Examination of the conjecture at the root of the tree lets us conclude that the robot is at the warehouse.

During the construction of a modified proof tree, each conjecture arising from the conjecture negation as a result of the addition of its negation to it is converted into tautology. Therefore, a conjecture located at the root logically follows from axioms and tautologies. But since the tautology may be omitted without affecting the incompletability or completability of any set containing other formulae (conjectures) besides tautologies, the indicated conjecture logically follows from axioms. Thus, the modified disproof tree is a graph (tree) of a proof that the formula located at its root logically follows from axioms.

Let us consider an example of the transformation into tautology of a more complex conjecture that arises during negation of a supposition. Let the supposition have the form

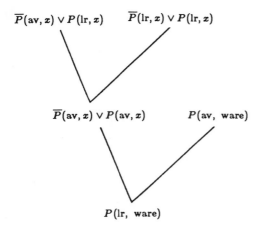

Figure 3.17 A modified proof tree in the loading robot problem.

$(\exists x \exists y)\{[F(x) \wedge G(x)] \vee [P(y) \wedge Q(y)]\}$

Its negation results in two conjectures

$\overline{F}(x) \vee \overline{G}(x); \ \overline{P}(y) \vee \overline{Q}(y)$

To transform the disproof graph into a modified proof tree, it is necessary to convert these conjectures into tautologies

$\overline{F}(x) \vee \overline{G}(x) \vee (F(x) \wedge G(x)); \ \overline{P}(y) \vee \overline{Q}(y) \vee (P(y) \wedge Q(y))$

However, these formulae are not conjectures since they contain conjunctions $F(x) \wedge G(x)$ and $P(y) \wedge Q(y)$. During the construction of a modified tree, they may formally be viewed as conjectures, and these conjunctions may be treated as single literals.

In general, the supposition has the form

$(\exists x_1) \ldots (\exists x_n)[F_1(x_1, \ldots, x_n) \vee \ldots \vee F_m(x_1, \ldots, x_n)]$

where each term F_i represents the conjunction of literals

$F_i = L_{i1} \wedge \ldots \wedge L_{ik_i}$

The negation of the supposition results in conjectures

$\overline{F}_i = \overline{L}_{i1} \vee \ldots \vee \overline{L}_{ik_i}, \quad i = 1, \ldots, m$

To convert them into tautologies, one must add to them respectively

$L_{i_1} \wedge \ldots \wedge L_{ik_i} = F_i$

Thus, at the root vertex of the modified graph we obtain the answer statement representing a disjunction of terms F_i, where instead of $x_1, \ldots, x_n$ there are other therms. It is obvious that each term $F_i(i = 1, \ldots, m)$ may or may not be included several times into the answer statement. Therefore, the answer statement (the formula at the root of the modified tree) does not necessarily have the same form as in the supposition.

If the modified proof tree is constructed without first constructing the disproof tree, then the resolutions are performed until the answer statement is obtained, i.e., the formula that represents a disjunction of only special cases of terms F_i contained in the conjecture.

Suppositions containing the commonality quantor. In the case when a supposition contains the commonality quantor, additional difficulties arise. This is because during negation, variables associated with the commonality quantor are transformed into variables associated with the existence quantor, and it becomes necessary to introduce the Skolem function. The difficulties are in the interpretation of these functions if they appear as therms in the answer statement. Let us demonstrate this by an example [1].

Suppose we are given a set of axioms

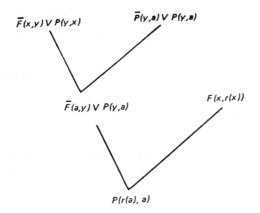

Figure 3.18   A modified proof tree with the Skolem function at its root.

$\Phi_0 = \{F(x, r(x)),\ \overline{F}(x, y) \vee P(y, x)\}$

where the following notation is used: $F(x, z)$, where x is a child of z; $r(x)$ is the function relating x to its parent $r(x)$; $P(y, x)$, where y is the parent of x.

Thus, we are given the following two axioms: 1) each x is a child of $r(x)$; 2) for all x and y, if x is a child of y, y is the parent of x

$$((F(x, y) \rightarrow P(y, x)) = \overline{F}(x, y) \vee P(y, x))$$

It is necessary to determine a parent for an arbitrary x. The supposition has the form $\forall x \exists y P(y, x)$. Transforming the negation of this supposition $\exists x \forall y \overline{P}(y, x)$ into the form of conjectures, we obtain $\overline{P}(y, a)$, where a is the Skolem function that does not contain arguments. A modified proof tree for this example is shown in Fig. 3.18. At the root, we have an answer statement $P(r(a), a)$ which is difficult to interpret, and which contains the Skolem function. The supposition negation $\overline{P}(y, a)$ means that some individual does not have parents. Therefore the answer statement should be interpreted as follows: regardless of an individual a, for whom there were not supposed to be parents, it turned out that $P(r(a), a)$ is valid. In other words, for any a, the parent is $r(a)$. It is clear that the cited proof would also be valid if a variable is used instead of a constant a.

It can be shown [1] that in the process of deriving an answer, we can replace any Skolem functions (and not just constants) arising during the negation of suppositions by new variables. During a proof, no substitutions into these new variables are performed. In the process of performing some resolutions, they may be renamed.

Automatic program writing. The described process of deriving an answer may be used for automatic construction of simple computer programs. Suppose we are given relationships $R(x, y)$ between x and y by some set of axioms, as well as elementary functions defined by another set of axioms. It is necessary to write a program that, on the basis of the given input variable, will output a value y that satisfies the relationship $R(x, y)$. In writing this program, the given elementary functions can be used.

The desired program may be constructed with the aid of the process of answer extraction after it is proven that the supposition $\forall x \exists y R(x, y)$ logically follows from the given axioms. At the root of the modified proof tree, we will have the answer statement y in the form of an elementary function composition. This composition will be the desired program.

Suppose that $R(x, y)$ defines the following relationship: x is an arbitrary list (vector) of numbers; y is an orderly arranged list containing the same numbers as x, but arranged in a descending order. Let us assume that in the program preparation we can use the following elementary functions:

$\operatorname{car}(x) : x = (x_1, \ldots, x_n) \rightarrow x_1$ (transforms list (vector) x into the first element of this list);

$\operatorname{cdr}(x) : x = (x_1, \ldots, x_n) \rightarrow (x_2, \ldots, x_n)$ (transforms list x in the same list without the first element);

$\operatorname{cons}(x_1, y) : \{x_1; \ y = (y_1, \ldots, y_n)\} \rightarrow (x_1, y_1, \ldots, y_n)$ (transforms element x_1 and list y into a new list obtained by placing x_1 before the list y);

merge (x_1, y) (transforms element x_1 and the ordered list y into an orderly arranged list containing elements of list y and element x_1).

Let us formulate the axioms that formalize the above definitions of the relationship and elementary functions. To do this, two additional relationships are introduced:

$S(y)$ — the list is orderly arranged;
$I(x, y)$ — two lists x and y consist of the same elements.

The relationship $R(x, y)$, with the aid of the introduced relationships $S(y)$ and $I(x, y)$, may be defined as follows

$$\forall x \forall y \{ \{ R(x, y) \rightarrow [S(y) \wedge I(x, y)] \} \wedge \{ [S(y) \wedge I(x, y)]$$

$$\rightarrow R(x, y) \} \}$$

Relationships $R(x, y)$, $S(y)$, and $I(x, y)$, with the aid of the elementary functions, are defined by the following formulae

$$R[\operatorname{cons}(\operatorname{car}(x), \operatorname{cdr}(x)), y] \rightarrow R(x, y)$$

$$\forall x \forall y \forall u \{ I(x, y) \rightarrow I[\operatorname{cons}(u, x), \operatorname{merge}(u, y)] \}$$

$I(\text{nil}, \text{nil})$(here "nil" means an empty list)

$$\forall x \forall y \{ S(y) \rightarrow S[\operatorname{merge}(x, y)] \}$$

$S(\text{nil})$

The combination of the above formulae serves as a formal (e.g., in the predicate calculus language) definition of the introduced relationships and of the elementary formulae. Let us transform it into a conjecture set

$$\overline{R}(x,y) \vee S(y), \ \ \overline{R}(x,y) \vee I(x,y), \ \ \overline{S}(y) \vee \overline{I}(x,y) \vee R(x,y)$$

$$\overline{R}[\text{cons}(\text{car}(x), \ \text{cdr}(x), y] \vee R(x,y)$$

$$\overline{I}(x,y) \vee I[\text{cons}(u,x), \ \text{merge}(u,y)]$$

$$I(\text{nil}, \ \text{nil})$$

$$\overline{S}(y) \vee S[\text{merge}(x,y)]$$

$$S(\text{nil})$$

The supposition has the form $\forall x \exists y R(x,y)$ and its negation has the form $\overline{R}(a,y)$, where a is the Skolem function.

The programming problem entails the expression of $y = \text{sort}(x)$ (sort is the ordering operator) by means of the above elementary functions.

Let us attempt to prove the supposition $\forall x \exists y R(x,y)$ by the method of mathematical induction: first for a list of zero length, and then, assuming that it is correct, for a list of the length $n \geq 0$; let us prove that it is valid for a list of the length $n + 1$.

If the lists are of a length $n = 0$, then the suppositions may be formulated in the form $\exists y R(\text{nil}, \ y)$. $\overline{R}(\text{nil}, y)$ will be a negation. After constructing a disproof tree, and then a modified proof tree, we will obtain the answer statement $R(\text{nil}, \ \text{nil})$. Thus, if the length of list x is equal to zero, then $y = \text{nil}$.

Suppose that for each nonempty list x, the value cdr (x) may be obtained by sorting (the induction supposition). This supposition may be written as $R(\text{cdr}(x), \ \text{sort}(\text{cdr}(x)))$ and may be included into the list of axioms. Having constructed a disproof tree and transformed it into a modified proof tree, we obtain the answer statement

$$R[x, \text{merge}[\text{car}(x), \ \text{sort}(\text{cdr}(x))]]]$$

where the Skolem function is excluded (replaced by a variable). Combining the results obtained for $n = 0$ and $n \neq 0$, we find

> sort $(x) = \text{nil}$, if $x = \text{nil}$
> sort $(x) = \text{merge} \ [\text{car} \ (x), \ \text{sort} \ (\text{cdr} \ (x))]$ — otherwise.

This relationship determines a recursive program of ordering a list of numbers of arbitrary length.

3.5   MAKING DECISIONS UNDER CONDITIONS OF LINGUISTIC UNCERTAINTY

In most cases, decision making in planning or management occurs under the conditions of uncertainty. Thus, in the operation of manufacturing facilities, downtime is inevitable (failure of individual elements or missed deliveries of required parts or components). Naturally, in the phase of design and development of a management system for such facilities, it is not known ahead of time which element will fail and when, and what parts will be in short supply. In this case, the uncertainty is due to random factors. Uncertainties of this kind will be called "probabilistic" or "stochastic" uncertainties. To formalize problems of decision making under the conditions of stochastic uncertainty, the probability theory is used, as well as the theory of statistical decisions and multiqueue server systems [5].

There are also possible uncertainties of another kind: uncertainties due to a blurred (diffused) purpose and/or constraints. Suppose, for instance, that we have to synthesize a control system for a transportation robot that must stop in the near vicinity of an obstruction, or that we have to develop a control algorithm for a robot that will reject cylindrical blanks if their length is significantly larger (or smaller) than the specified length. In these examples, the uncertainty stems from the lack of clarity of such concepts as "near," "significantly larger," "significantly smaller." Such uncertainties are termed as "linguistic" uncertainties [6]. A linguistic uncertainty is always present in the so-called humanistic systems, that is, systems whose behavior is strongly affected by opinions, perceptions, or emotions of a human.

To formalize problems of decision making with unclear purpose and/or constraints of the construction of a mathematical model of systems that contain linguistic uncertainties, a theory of unclear or diffuse sets has been developed.

Elements of the theory of diffuse sets. Suppose we have a universal set $\mathbf{X}$ designated as space. It is intuitively clear that a diffused set of elements of space $\mathbf{X}$ is a set that does not have a strictly-defined boundary between its elements and other elements $\mathbf{X}$.

Let $\mathbf{X}$ be a set of some parts of various dimensions, let $\mathbf{A}$ and $\mathbf{B}$ be sets of large and small parts from $\mathbf{X}$, respectively. It is obvious that $\mathbf{A}$ and $\mathbf{B}$ are examples of diffused sets in $\mathbf{X}$.

Main concepts and definitions. A nondiffused set $\mathbf{A}$ is defined as a combination of orderly arranged pairs

$$\mathbf{A} = \{x, \chi_{\mathbf{A}}(x)\}, \ x \in \mathbf{X}$$

where $\chi_{\mathbf{A}} : \mathbf{X} \rightarrow \{0, 1\}$ is the characteristic function with

$$\chi_{\mathbf{A}}(x) = \begin{cases} 1, x \in \mathbf{A} \\ 0, x \notin \mathbf{A} \end{cases}$$

The definition of a diffused set is obtained as a generalization of this definition. A "diffused set" $\mathbf{A}$ in $\mathbf{X}$ is a combination of orderly arranged pairs

$$\mathbf{A} = \{x, \mu_A(x)\}, \ x \in \mathbf{X} \tag{3.7}$$

where $\mu_A : \mathbf{X} \rightarrow [0, 1]$.

The function μ_A is called an "affiliation function," and its value $\mu_A(x)$ is called the "degree of affiliation" with $\mathbf{X}$. In a special case when $\mu_A(x)$ assumes only values 0 or 1, set 3.7 is nondiffused. If $\forall x \in \mathbf{X} \ \mu_A(x) = 0$, then $\mathbf{A}$ is an empty set $(\mathbf{A} \neq \emptyset)$; if $\forall x \in \mathbf{X} \mu_A(x) = 1$, then $\mathbf{A}$ coincides with the entire space $\mathbf{X}(\mathbf{A} = \mathbf{X})$.

A diffused set $\mathbf{A}$ is called "normal" if $\sup_{x \in \mathbf{X}} \mu_A(x) = 1$; a non-normal set is called "subnormal." A nonempty set $\mathbf{A}$ may always be normalized by dividing the affiliation function by $\sup_{x \in \mathbf{X}} \mu_A(x)$, i.e., be made a normal one.

A set $\mathbf{S(A)} = \{x : \mu_A(x) > 0, x \in \mathbf{X}\}$ is called a "carrier" of the set $\mathbf{A}$. In a special case, when $\mathbf{A}$ is a conventional set, the carrier coincides with the set itself $(\mathbf{S(A)} = \mathbf{A})$.

In a class (in a set) of diffused sets, there are relationships and operations analogous to those used in the class of conventional sets. Below, speaking about a set, we will mean a diffused set. A conventional set is treated as a special case of a diffused set.

Two sets $\mathbf{A}$ and $\mathbf{B}$ are equal (it is written $\mathbf{A} = \mathbf{B}$) if $\mu_A = \mu_B$ (i.e., $\forall x \in \mathbf{X} \mu_A(x) = \mu_B(x)$). A set $\mathbf{A}$ is contained in the set $\mathbf{B}$ (or $\mathbf{B}$ includes $\mathbf{A}$) if $\mu_A \leq \mu_B$ (i.e., $\forall x \in \mathbf{X} \mu_A, (x) \leq \mu_B(x)$). A set $\mathbf{A'}$ (sometimes denoted by $]\mathbf{A}$ or $\overline{\mathbf{A}}'$ is a compliment to set $\mathbf{A}$ if $\mu_{A'} = 1] - \mu_A$.

An "intersection" of two sets $\mathbf{A}$ and $\mathbf{B}$ (denoted by $\mathbf{A} \cap \mathbf{B}$) is a set $\{x, \mu_{A \cup B}(x)\}$ where

$$\mu_{A \cup B}(x) = \min\{\mu_A(x), \ \mu_B(x)\}, \ x \in \mathbf{X} \tag{3.8}$$

A "combination" of sets $\mathbf{A}$ and $\mathbf{B}$ (denoted by $\mathbf{A} \cup \mathbf{B}$) is a set $\{x, \mu_{A \cup B}(x)\}$, where

$$\mu_{A \cup B}(x) = \max\{\mu_A(x), \ \mu_B(x)\} \tag{3.9}$$

Using the signs of conjunction $\wedge$ and disjunction $\vee$, the last two relationships are also written as

$$\mu_{A \cup B} = \mu_A \wedge \mu_B \tag{3.10}$$

$$\mu_{A \cup B} = \mu_A \vee \mu_B \tag{3.11}$$

It should be remembered that (3.10) and (3.11) are the conditional records of (3.8) and (3.9), respectively.

An algebraic product of sets $\mathbf{A}$ and $\mathbf{B}$ (denoted by $\mathbf{AB}$) is a set $\{x, \mu_{AB}(x)\}$, where

$\mu_{AB}(x) = \mu_A(x)\mu_B(x), \quad x \in \mathbf{X}$

An algebraic sum of sets $\mathbf{A}$ and $\mathbf{B}$ (denoted by $(\mathbf{A} \oplus \mathbf{B})$) is a set $\{x, \mu_{A \oplus B}(x)\}$, where

$\mu_{A \oplus B}(x) = \mu_A(x) + \mu_B(x) - \mu_A(x)\mu_B(x)$

Operations $\cap$ (intersection) and $\cup$ (combination), as can be easily verified, are associative and distributive with respect to one another, i.e., if $\mathbf{A}$, $\mathbf{B}$, $\mathbf{C}$ are arbitrary sets in $\mathbf{X}$, then

$\mathbf{A} \cap \mathbf{B} \cap \mathbf{C} = (\mathbf{A} \cap \mathbf{B}) \cap \mathbf{C} = \mathbf{A} \cap (\mathbf{B} \cap \mathbf{C})$

$\mathbf{A} \cup \mathbf{B} \cup \mathbf{C} = (\mathbf{A} \cup \mathbf{B}) \cup \mathbf{C} = \mathbf{A} \cup (\mathbf{B} \cup \mathbf{C})$

$(\mathbf{A} \cap \mathbf{B}) \cup \mathbf{C} = (\mathbf{A} \cup \mathbf{C}) \cap (\mathbf{B} \cup \mathbf{C})$

$(\mathbf{A} \cup \mathbf{B}) \cap \mathbf{C} = (\mathbf{A} \cap \mathbf{C}) \cup (\mathbf{B} \cap \mathbf{C})$

Operations of algebraic product and algebraic sum are associative but not distributive.

A binary diffused relationship P in $\mathbf{X}$ is defined as a diffused set in a direct product $\mathbf{X}_1 \times \mathbf{X}_2$ (here $\mathbf{X}_1$ and $\mathbf{X}_2$ are different specimens of space $\mathbf{X}$)

$P = \{(x_1, x_2), \ \mu_P(x_1, x_2)\}, \quad x_1 \in \mathbf{X}_1, \ x_2 \in \mathbf{X}_2$

where $\mu_P(x_1, x_2)$ is the degree of affiliation (x_1, x_2) to P.

In general, the n-th order diffuse relationship P_n in $\mathbf{X}$ is a diffused set in a direct product $\mathbf{X}_1 \times \ldots \times \mathbf{X}_n$

$P_n = \{(x_1, \ldots, x_n), \ \mu_{P_n}(x_1, \ldots, x_n)\}$

Let $\mathbf{X} = R$, where $R = (-\infty, \infty)$, to be a real line. The condition $x_1 \gg x_2$ (x_1 is much greater than x_2) gives a diffused relationship in R, whose affiliation function may [7] have the form

$$\mu_P(x_1, x_2) = \begin{cases} 0, x_1 \le x_2; \\ \left[1 + (x_1 - x_2)^{-2}\right]^{-1}, & x_1 > x_2 \end{cases} \tag{3.12}$$

A diffused set $\mathbf{C}$ in space $\mathbf{X} \times \mathbf{Y}(\mathbf{X} = \{x\}, \ \mathbf{Y} = \{y\} = \mathbf{Z}$ with the affiliation function $\mu_C(x, y)$ is called "expendable" by $\mathbf{X}$ and $\mathbf{Y}$ if $\mathbf{C}$ permits the representation $\mathbf{C} = \mathbf{A} \cap \mathbf{B}$ or, which is the same

$\mu_C(x, y) = \mu_A(x) \wedge \mu_B(y)$

where $\mathbf{A}$ and $\mathbf{B}$ are diffused sets in $\mathbf{X}$ and $\mathbf{Y}$ with the affiliation functions $\mu_A(x)$ and $\mu_B(y)$, respectively.

Diffused sets generated by transformations. Let $f : \mathbf{X} \rightarrow \mathbf{Y}$ be a transformation (nondiffused) from $\mathbf{X} = \{x\}$ into $\mathbf{Y} = \{y\}$. Then $y = f(x)$ is called a "transform" ("image") of element x; $x \in \mathbf{X}$ is called "inverse image" of y if $f(y) = y$. A set of all inverse images of y is denoted by $f^{-1}(y)$.

If $\mathbf{A}$ is a diffused set in $\mathbf{X}$ with the affiliation function μ_A, then its image during transformation of f is called a diffused set $\mathbf{B} = f\,(\mathbf{A})$ with the affiliation function determined by the following relationship

$$\mu_B(y) = \begin{cases} \sup_{x \in f^{-r}(y)} \mu_A(x), \ f^{-1}(y) \neq \emptyset \\ 0, f^{-1}(y) = \emptyset \end{cases}$$

If $\mathbf{B}$ is a diffused set with the affiliation function μ_B, then the inverse image of this set during transformation of f is called a diffused set $\mathbf{A} = f^{-1}\,(\mathbf{B})$ with the affiliation function $\mu_A(x) = \mu_B(f(x)) \forall x \in \mathbf{X}$.

Let $\mathbf{B}\,(x)$ be a diffused set in space $\mathbf{Y} = \{y\}$, the affiliation function of which depends on variable x as a parameter, which is written as $\mu_B(y/x)$. Then the set $\mathbf{B}\,(x)$ is called "conditional" with respect to x or simply "conditional diffused set." Let us assume that the region of variation of parameter x is space $\mathbf{X}$ and to every $x \in \mathbf{X}$ there corresponds a diffused set $\mathbf{B}\,(x)$ in $\mathbf{Y}$. In this case, function $\mu_B(y/x)$ determines the transformation from $\mathbf{X}$ into a space of diffused sets in $\mathbf{Y}$. Such a transformation is called a "diffused (or "blurred") transformation" from $\mathbf{X}$ into $\mathbf{Y}$. In other words, a transformation $\mathbf{B}: \mathbf{X} \rightarrow \mathbf{Y}$ that assigns to each element x from $\mathbf{X}$ a diffused set $\mathbf{B}\,(x)$ in $\mathbf{Y}$ is called a "diffused (or "blurred") transformation" from $\mathbf{X}$ into $\mathbf{Y}$.

A transform of a diffused set $\mathbf{A}$ into $\mathbf{X}$ for a diffused transformation with the affiliation function $\mu_B(y/x)$ is called a diffused set $\mathbf{B}$ defined by the affiliation function

$$\mu_B(y) = \sup_{x \in \mathbf{X}} \min\{\mu_A(x),\ \mu_B(y/x)\}$$

A diffused solution [7]. Let $\mathbf{X} = \{x\}$ be the space of alternatives (i.e., possible solutions). It is desirable to find (select) a solution for an unclear purpose $\mathbf{G}$ and blurred constraint $\mathbf{C}$. In formalizing this problem within the framework of the theory of diffused sets, the unclear purpose $\mathbf{G}$ is identified with a fixed diffused set $\mathbf{G}$ in $\mathbf{X}$, while the blurred constraint $\mathbf{C}$ is identified with some diffused set $\mathbf{C}$ in $\mathbf{X}$. In other words, both the purpose and the constraint are represented as diffused sets; this gives one the basis not to differentiate between them in the selection of a solution.

Suppose that from a given set of parts we have to selectively sort those part whose mass is considerably higher, say 2 kg (purpose $\mathbf{G}$), and is approximately in the range between 8 and 10 kg (constraint $\mathbf{C}$). In formalizing this problem, the purpose may be identified with a diffused set $\mathbf{G}$ defined by the affiliation function of the form, for instance see (3.12)

Table 3.1

x	5	6	7	8	9	10	11	12
μ_G	0.900	0.961	0.973	0.980	0.985	0.988	0.990	
μ_C	0.0376	0.110	0.385	0.909	1.0	0.909	0.385	0.110
μ_D	0.0376	0.110	0.385	0.909	0.980	0.909	0.385	0.110

$$u_G(x) = \begin{cases} 0, x \leq 2 \\ \left[1 + (x - 2)^{-2}\right]^{-1}, & x > 2 \end{cases}$$

and the restriction may be identified by a diffused set **C** defined by the affiliation function

$$\mu_C(x) = [1 + 0, 1(x - 9)^4]^{-1}$$

Let us determine the solution for diffused purposes and constraints. Suppose in the space of alternatives **X** we are given a diffused purpose **G** and a diffused constraint **C**. Then the solution is determined as an intersection of **G** and **C**. In general, when there are n diffused purposes $(G_i(i = 1, \ldots, n)$ and m diffused constraints $C_j(j = 1, \ldots, m)$, the solution **D** is determined as an intersection of all given purposes and constraints

$$\mathbf{D} = \mathbf{G}_1 \cap \ldots \cap \mathbf{G}_n \cap \mathbf{C}_1 \cap \ldots \mathbf{C}_m$$

$$\mu_D = \mu_{G_1} \wedge \ldots \wedge \mu_{G_n} \wedge \mu_{C_1} \wedge \ldots \wedge \mu_{C_m}$$

Thus, a solution for diffused conditions and purposes is a diffused set and is called a diffused solution, which may be viewed as an unclearly formulated instruction, conditioned by unclear purposes and constraints. A variety of approaches to performing these instructions is possible. One reasonable approach is the selection of such an option from **D** that would have the maximum degree of affiliation with **D**. A combination of alternatives (solutions) from **D** sometimes is called the "optimum" solution, and each such alternative (i.e., any element from this combination) is called a maximizing solution. If the maximizing solution is unique, then obviously the maximizing and optimum solutions coincide.

Suppose we wish to find a solution of a problem of sorting parts in accordance to their weight as described in the previous example. Table 3.1 lists values of the affiliation functions μ_G, μ_C, and μ_D.

For simplicity, let us consider only the integer values of x. It can be seen from the Table that in the integer class of solutions the only maximizing solution is $x^* = 9$.

In the above definition of a diffused solution, it is assumed that all given purposes and constraints have equal importance. However, there are possible cases when some purposes and restrictions may be more important than others. In these cases, the solution may be defined as a diffused set $\mathbf{D}$ with the affiliation function of the form

$$\mu_D(x) = \sum_{i=1}^{n} a_i(x)\mu_G(x) + \sum_{j=1}^{m} \beta_j(x)\mu_{C_i}(x)$$

where $a_i(x), \beta_j(x)$ are weight functions that satisfy the following constraint

$$\sum_{i=1}^{n} a_i(x) + \sum_{j=1}^{m} \beta_j(x) = 1$$

Weight functions may be selected in such a way as to take into the account the relative importance of purposes and constraints.

Previously we considered situations when purposes and constraints were diffused sets in the space of alternatives. In a more general case, purposes and constraints may represent diffused sets in different spaces.

Let us suppose that restrictions are given as diffused sets $\mathbf{C}_1, \ldots, \mathbf{C}_m$ in the space of alternatives $\mathbf{X}$, while purposes are given as diffused sets $\mathbf{G}_1, \ldots, \mathbf{G}_n$ in some other space $\mathbf{Y}$. In addition, we are given the transformation $f : \mathbf{X} \to \mathbf{Y}$. In this case, purposes $\mathbf{G}_i(i = 1, \ldots, n)$ may be transformed and presented in the form of diffused sets $\overline{\mathbf{G}}_i(i = 1, \ldots, n)$ in $\mathbf{X}$ which are inverse images of sets $\mathbf{G}_i(i = 1, \ldots, n)$ of the transformation f, i.e., $\overline{\mathbf{G}}_i = f^{-1}(\mathbf{G}_i)$ and $\mu_{\overline{G}_i}(x) = \mu_{G_i}(f(x))(i = 1, \ldots, n)$. Then the affiliation function of solution $\mathbf{D}$ is defined by the equality

$$\mu_D(x) = \mu_{G_i}(f(x)) \wedge \ldots \wedge \mu_{G_n}(f(x)) \wedge \mu_{C_1}(x) \wedge \ldots \wedge \mu_{C_m}(x) \qquad (3.13)$$

Thus, cases when purposes and restrictions are given as diffused sets in different spaces are reduced to a case when they are given as diffused sets in the same space (space of alternatives).

FOUR
COMMUNICATION WITH COMPUTER AND MACHINE PERCEPTION

Growth in the number of computers and a rapid increase in the usage of mathematical methods and automatic systems have brought about a strong "computer dependence" of professional activity of specialists in such areas as design, planning, technology management, and so on. Direct interaction between a specialist and a computer is made difficult by the "language barrier," i.e., a difference between the professional language of the user and communication languages of a computer. The solution of this problem lies in the creation of intelligent communication means permitting dialogue with a computer to be conducted in a professional language. Another important problem, the solution of which ensures direct interaction of a computer with the outside world, is the development of a system of visual perception.

4.1 LANGUAGE COMMUNICATION SYSTEMS

Language communications systems may be divided into four groups: formatted, text-based, with limited logic, and with general derivation.

Formatted systems are among the simplest dialogue systems and commonly use two rigid formats: one, for knowledge representation stored in the system; and the other, for representing input and output information. A communication model that uses a tabular language is an example. An input information is a table filled out by a human. An output message may also be in the form of a table. A table rigidly controls the format and order of words in sentences, which simplifies software and methods for storing data in a computer. Formatted systems are employed, for instance, in developing simple information-reference and accounting systems. They have a limited purpose and are characterized by information rigidity which does not permit any manipulations that could make it usable for other purposes.

In text-based systems, text is stored directly in a natural language. Text is equipped with a variety of indexing schemes that permit data to be requested and searches of requested sentences or text fragments to be performed. The most widely used systems in this group are those which use descriptive languages. Descriptors are lexicon units of informational-search language that have a unique meaning. They are used to index text. Descriptors and relations among them are established on the basis of an analysis of the professional user language and are formalized in the form of a descriptor graph. During communications, use of words from the descriptor dictionary is permitted. The structure of a request is constructed in accordance with the communication structure among the descriptors. For instance, if the system is designed as a reference for a computer specialist, then the following words may serve as descriptors: COMPUTER, MINI, SUPER, PERSONAL, PROCESSOR, SIZE, MEMORY, etc. A descriptor language may be viewed as a limited model of a professional language. The disadvantage of these languages is that it is impossible to organize more complex selection procedures than a search using the descriptor language.

Systems with limited logic utilize formal knowledge representation models with simplified systems of logical derivation. Among the better known systems within this group are those implemented with the use of such languages as PLANNER, LISP, and PROGRAMMER (a modification of LISP).

Systems with general derivation utilize a broad class of formal models such as semantic networks, frame networks, blurred linguistic variables, predicate models, etc. In a number of cases, communication systems are implemented by using the universal procedure of theorem proof. The form of information representation in these systems does not depend on the features of the communication system. This property permits the use of such systems in any area that can be represented in the predicate calculus. However, the true-life complexity of the natural language is difficult to represent in binary logic, which is the disadvantage of predicate models.

It is obvious that an intelligent (from the point of view of language communication) system must understand text in natural or limited natural language.

To understand text means to transform it into its internal representation and link it with the base of knowledge. The process of this transformation is divided into three phases: morphological analysis, syntactic analysis, and semantic analysis.

As a result of the morphological analysis, in each work of the input-sentence a base is isolated and morphological information assigned to it, which permits the word to be assigned to a specific syntactic group and relationships among the words that are necessary for a syntactic analysis to be established. For a morphological analysis, a dictionary of input words and morphological tables containing word bases and lists of affixes (suffixes and endings) with comments is required. A block diagram of the morphological analysis process is shown in Fig. 4.1. First one verifies that a word belongs to the class of invariant words. The lexical unit to which an invariant word belongs may be determined right away. In the word does not belong to the class of invariant words, then a search of its base and affixes is performed. The morphological information is formed next. At the output of the morphological block, a sequence of conditional units is obtained, each of which is

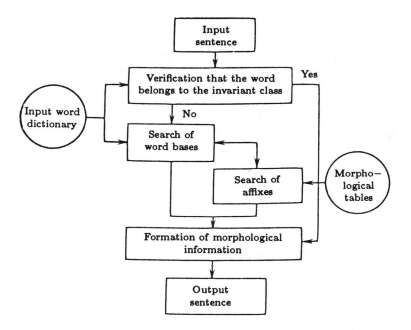

Figure 4.1   Block diagram of the morphological analysis process.

compared with the input-sentence word and which contains information about its affiliation with a specific lexical unit.

The purpose of "syntactic analysis" consists of the construction of a structure of an input sentence determined by the specific language grammar. In recording sentences in the language defined by a context-free grammar, the purpose of syntactic analysis consists of obtaining the component tree. In processing a language described by transformational grammar, syntactic analysis boils down to obtaining the profound sentence structure. One of the effective methods of obtaining the profound structure is the method of expanded transition networks.

Let us first examine a transition network with a finite number of states. It is a graph, whose vertices (states) correspond to nonterminal language symbols, while its arcs correspond to terminal language symbols. We identify states marked by an initial symbol S of the grammar, and final states are marked by S^*. An example of such a network for a subset of Russian-language sentences is shown in Fig. 4.2. The following notation is used in Fig. 4.2: N — noun, ADJ — adjective, V — verb, PR — preposition. To generate some chain on the basis of the transition network, the following rules are used:*

1. Generation begins at vertex S.

*Translator's note. In an attempt to keep as close to the text as possible, there is some slight discrepancy between the test and the figure.

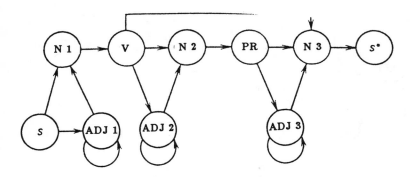

Figure 4.2   Transition network with a finite number of states.

2. A step is made along one of the arcs into the next state. A word that is next in sequence is read and verified, whether or not this word is present in the nonterminal words dictionary and whether or not it can become a marker for one of the arcs originating from a given state.

3. The generation process stops if vertex S^* is reached. The arc leading to this vertex is marked by a symbol $*$. The obtained chain of terminal symbols comprises the generated sentence.

Suppose, for instance, that we wish to obtain a sentence using the network of Fig. 4.2 that reads: LARGE POWERFUL CRANE LOWERS A CONTAINER ON A PLATFORM. It is obvious that the order of examining the nodes will be as follows: S, ADJ1, ADJ1, N1, V, N2, PR, N3, S^*. Using this network we can also generate another sentence: A LOCOMOTIVE PULLS HEAVY TRAIN ON A RAILROAD. The order of examining nodes is: S, N1, V, ADJ2, N2, PR, ADJ3, N3, S^*. A number of other sentences can be generated in a similar manner.

Networks with a finite number of states have limited capabilities. For instance, with their aid one cannot model languages that include rules of recursive generation, i.e., rules of the form $A \rightarrow \varphi_1 A \varphi_2$. A recursion may be set up by expanding the model of a network with a finite number of states in the following manner. Suppose a grammar is represented by a set of networks with a finite number of states, each of which corresponds to a grammatical analysis of certain terminal language components. Isolated vertices in the networks are marked by special symbols. During a transition of the process into a state corresponding to a marked vertex, another network is called up (the control is turned over to the called-up network). The network which turns over control is called a "calling" network. The network taking over control is called a "called" network. The called network may call up other networks. Also possible are recursive calls, when a network returns to its initial state either directly or via an intermediate network.

A system of transition networks with a finite number of states and calling procedure is called a "recursive network." With the aid of recursive networks, it is possible to model any context-free grammar.

A recursive network, just like a network with a finite number of states, permits a superficial, rather than profound sentence structure, to be uncovered. Further expansions of transition networks permit transformations to be performed and profound sentence structure to be uncovered.

An expansion of a recursive network entails the following:

- arcs are assigned transition conditions, i.e., a transition from state to state is performed only after meeting these conditions;

- arcs are assigned operations that are performed if a transition along a given arc takes place.

A network with such properties is called an "expanded transition network" and is used to construct partial structural description of a sentence. Description fragments are stored in registers organized in stacks. Operations involving arcs change the register content and the order of examining the network sectors. Information contained in registers may also be used to redefine the input sentence structure. It is possible to perform in the software trial assignments of parts of an input chain to various sentence structures. If subsequent analysis shows that such an assignment is incorrect, then it is nullified.

Input information in a semantic analysis is a profound structure of an input sentence. The problem of the semantic analysis is in understanding the meaning of a sentence. More precisely, the problem of understanding may be defined in the following manner [5].

- to formulate a unique representation of an input sentence in terms of the base knowledge;

- to combine an input sentence with the base knowledge.

A semantic interpretation must be based on an internal model of an object domain represented by the knowledge base. It can be easily shown that a syntactically tame phrase may not be semantically tame. For instance, phrases "A robot picked up a part" and "A part picked up the robot" are syntactically tame. The second phrase, however, has no meaning in such an object domain as the robot universe.

One major difficulty in semantic analysis is overcoming the multiple meaning of phrases of the natural language. Let us examine the elimination of ambiguity based on the introduction of additional syntactic categories. Additional syntactic categories expand the class of conventional categories (noun, verb, adjective, etc.) by introducing such terms as "animated," "nonanimated," "abstract," etc. In general, it is possible to construct a multilevel structure of semantic categories. Interpretation programs that use this method perform a sentence analysis from the "bottom up." First, category markers are assigned to each word, then, on the basis of these markers, markers for categories of higher vertices of the analysis tree are formulated. Using the procedures of category comparison, one succeeds in identifying meaningless word combinations while selecting an interpretation. However, the method of semantic categories is limited. It is impossible with its aid to uncover the ambiguity that arises from insufficient information within a sentence. Consider, for instance, a sentence: "A robocar arrived at the platform with parts." The location of the parts (in the robocar or on the platform) may be determined only on the basis of an analysis of previous states of the model of the object region. Therefore, in general, the discovery of ambiguity consists of the composition of all possible interpretations with the knowledge base. Several principles of such a comparison may be pointed out, such as [1]:

1. An interpretation that corresponds to a part of the database with the most connections among the vertices is selected.

2. An interpretation is selected such that, by combining it with the knowledge base, a minimum number of vertices and arcs is added.

3. Some measure of heuristic nature is used during the selection.

4. All interpretations are memorized and operations are formulated to obtain new information for eliminating ambiguity. Such an operation may be, for instance, a request of specific information from the user.

One variety of the semantic analysis is based upon the representation of an input sentence in the canonical form. The canonical representation is obtained as a result of a treatment of a limited natural language by an expanded transition network. The basis of the representation is the identification of superficial and canonical verbs. A group of superficial verbs describing one and the same operation (situation) is made to correspond to one profound canonical verb. In the dictionary, for every superficial verb there is a reference to the so-called A-rules and its canonical verb. A-rules contain information concerning prepositions and nouns used in conjunction with this verb for expressing various profound relationships and for indicating semantic categories to which this noun belongs.

For instance, superficial verbs TO BUY, TO ACQUIRE, correspond to a canonical verb TO EXCHANGE. Profound relationships are determined by the following nouns: BUYER, SELLER, OBTAINED, HANDED OVER, TIME, PLACE. An input sentence: "Plant A purchased from business B a machine tool costing N rubles" in the canonical form will be presented as follows: PURCHASER (plant A), SELLER (business B), OBTAINED (machine tool), HANDED OVER (N rubles).

After performing a canonical representation of a sentence, procedures are activated which make changes in the knowledge base. One method for making changes is as follows. A canonical verb is viewed as a procedural event, i.e., as an operator that transforms the system from one state into another. In case of knowledge representation by a semantic network, the result of performance of a given operator are structural changes in the network.

In particular, in the example with the machine tool purchase, a change will consist of the removal of the relationship PROPERTY between the vertices PLANT and MACHINE TOOL and placement of this relationship between the vertices BUSINESS and MACHINE TOOL.

Systems with general derivation have gained prominence in intelligent interfaces. The interface is a software package providing a direct dialogue between an end-user and a computer. An end-user is a specialist that formulates a problem in terms of an object region and that uses the results of its computer solution.

To understand the role of an intelligent interface in end-user-oriented dialogue systems, let us consider the traditional process of formulating and solving a problem on a computer (a problem here, as before, is understood in a broad sense). It may be divided into three procedures. The *first procedure* consists of problem formulation in terms of the object domain, i.e., in the professional language of the end-user. Such a formulation is often called "meaningful" since it most completely and precisely reflects the essence of the problem at hand and the requirements to the results of the solution. The *second procedure* is a mathematical statement of the problem, i.e., its formulation in the language of a mathematician. During its realization, it is necessary to make a transition from a nonformalized language of the user to a rigorous formal statement of the problem with the utilization of, for instance, such categories as a target function, constraints, a system of equations, etc. The *third procedure* is the programming of the problem. More often than not a program is written in a high-level language or in assembler. The *fourth procedure* is the representation of the problem in an internal language of the computer, i.e., in the form of data and commands used by a computer while solving the problem.

The realization of the described sequence of procedures may be presented by a diagram, shown in Fig. 4.3. Three phases of the translation that implement the transformation of forms of problem representation are identified in the diagram. The last (third) translation phase is performed by means of programs making up the computer software. The first and the second translation phases are performed by the appropriate specialists.

The presence of the "middle men" between the end-user and the computer increases the problem solution time and also increases the risk of distorting the meaningful problem statement during the first two translation phases because of the differences in the professional orientation and languages of the user, the mathematician, and the programmer. The main purpose of an intelligent interface is to eliminate the middle men between the end-user and the computer.

In general, the structure of an intelligent interface consists of four functional components: a semantic model, means of logical derivation, a linguistic processor, and a planner.

The semantic model is the main component and is constructed using one of the systems of knowledge representation described in Chapter 2.

The means of logical derivation are used both for solving logical problems and for implementing procedures for supplementing the knowledge base.

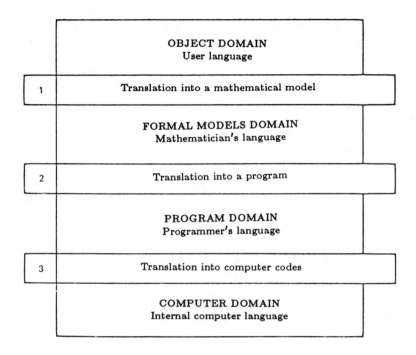

Figure 4.3   Diagram of statement and solution of problem on a computer.

The linguistic processor provides communications of the end-user with the computer in the professional dialect of a natural language. The operation of the linguistic processor is based on language models in the object domain of the user. In general, it performs the described phases of transforming a language text into an internal representation, i.e., the morphological, the syntactic, and the semantic analysis.

The planner formulates the final problem solution program based on the initial description prepared by the end-user. In doing this, unified functional modules assembled in a special library are employed.

An intelligent interface is necessary in question-answer, computational-logical, and expert systems. However, their specificity affects the structure of an intelligent interface and the implementation method of its components.

Let us examine more closely an intelligent interface using an example of computational-logical systems, i.e., systems permitting the solution of computational problems on the basis of their statement and description in terms of the object domain of the end-user.

A semantic model of an object domain most often is constructed in the form of a semantic or a frame network. In the case of a semantic network, vertices are made to correspond to parameters that, depending on the problem statement, may be either initial data or values that have to be computed (unknowns). Network arcs are assigned with relationships of a mathematical type, i.e., formulae, operators, or more complex mathematical expressions transforming one parameter into another. To each mathematical network relationship, there is a corresponding computational module from the module library. From a semantic network of this type, it is possible to obtain solution algorithms of various computational problems in the form of an oriented network subgraph.

The interaction with the user in computational-logical systems takes place in the form of a dialogue conducted in the professional language of the user and is maintained by the linguistic processor. The internal representation of the problem is constructed during the dialogue and its analysis is performed by a special program. As a result of the analysis, initial data and unknowns are identified and assigned to corresponding vertices of the semantic network. The planner constructs all paths leading from the "initial-data" vertices to the "unknown" vertices. The oriented subgraph derived in this manner is the algorithm of the problem solution. In accordance with its structure, specific program modules are called from the library and implemented. Solution results are submitted to the user with the aid of the linguistic processor.

4.2 PERCEPTION OF VISUAL INFORMATION

In psychology, the term "perception" signifies a process of transforming actual objects and events into combinations of their various properties and parts associated with the understanding of the entirety of the reflected information. Unlike sensations that transform only separate properties of objects, the perception is integral in nature, i.e., it combines separate sensations into an integrated image.

To define machine perception, let us consider the concepts of an image and representation. An "image" is an array of events combined by common properties. Familiarization with one of these events permits one to make a judgement about the entire array of these events. Each specific event from the array of those combined into an image is called a "representation." For instance, an array of different planar representations of a rectangle may be termed as an image of a "rectangle." Properties of an image are a closed figure formed by four straight lines intersecting at right angles. Specific representations of a rectangle may differ in absolute and relative dimensions of the sides and location in space.

A machine perception is an information process aimed at the construction of an object domain model, i.e., at the creation of an internal machine representation of an integrated image for its subsequent purposeful analysis.

Perception is characterized by two important properties [4]: "constancy" and selectivity. The constancy is the ability to retain a received integrated image in the presence of variations in the object representation that are associated with variations in the distance, observation point, lighting conditions, etc. The constancy of perception may be used to characterize the power of a perception system. Selectivity is the ability of a system to isolate only those image details that are significant for the considered problem. The selectivity characterizes the system specialization and is specified during its design phase.

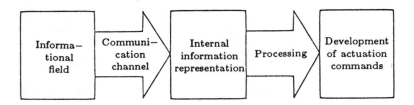

Figure 4.4 Diagram of information transformation in perception systems.

A general diagram of a process of information transformation in perception systems is shown in Fig. 4.4. The term "information field" as used here designates that part of object domain properties which is accessible by the sensing system. The communication channel is a combination of information processes of transmission and preliminary information processing for converting it into the internal representation, i.e., into a model of the perceived part of the object region.

Internal representation processing is performed by taking into account the system purpose and the type of the problem it is solving. One may identify two types of perception purposes: image recognition and scene analysis. The purpose of image recognition is to classify a specific representation as belonging to a specific image. Scene analysis is a broad class of problems associated with the perception of a combination of three-dimensional objects.

Depending on the nature of the problems at hand, perception systems are divided into three classes.

1. Systems of complete perception. In these systems, the transformation of elements of an internal field into elements of the internal representation is characterized by a one-to-one correspondence. In other words, all initial information must be submitted to the system. The main task of such systems is undistorted and complete perception of information. Systems of this kind include, for example, the majority of systems for transmitting, storing, and processing video information.

2. Systems of limited perception. Within these systems, multiple transformation of the information field into the internal representation is performed. This means that several elements of the information field are made to correspond to a single element of the internal representation. Information perceived by such a system is incomplete and ambiguous. Most machine vision systems in industrial robotics belong to the second class because they do not require precise copying of the environment. The internal representation should have the necessary completeness only so far as being sufficient for solving the orientation and manipulation problems. In a number of practical problems, a binary internal object representation turns out to be sufficient, i.e., a representation based on only two levels of brightness. The information volume processed by such systems is substantially smaller than in systems of the first class.

3. Systems of sequential perception. These systems are characterized by sequential analysis of the information field elements. Here all elements of the internal representation are made to correspond to sequentially examined groups of the information field elements.

At the present time, there are three main arrangements of internal representation of video information: two-dimensional transformations, syntactic methods, and recursive structures [4].

Fourier, Adamar, and other transforms are used most often as two-dimensional transformations, which permit signs (information characteristics) of a representation to be obtained. The advantage of such representation is the possibility to retain complete information about the representation. However, two-dimensional transformation is not effective in cases when one has to identify and analyze within a representation of objects that are hierarchically related among each other[†].

Syntactic methods presuppose the identification of structural representation elements which are viewed as terminal symbols of some formal language. Here one possible representation is the representation by a graph whose vertices are identified with the representation elements and arcs (with relationships among them). Algorithms constructed on the basis of the syntactic approach permit the solution of recognition problems with a sufficiently large number of classes.

Representations based on recursive structures also presuppose representation decomposition and the establishment of relationships between the elements. However, unlike the syntactic methods, here not the structural representation elements are identified, but rather the brightness characteristics of groups of representation elements. Regular representation decomposition (for instance as pyramid structures) permit the reconstruction of the entire representation, as well as the examination of its arbitrary approximations.

Let us list, in the order of increasing complexity, typical tasks for visual perception characteristic for robotics and industrial automated systems.

1. The determination of planar coordinates of known objects.

2. Recognition of isolated two-dimensional objects.

3. Recognition of touching two-dimensional objects.

4. Recognition of isolated three-dimensional objects.

5. Recognition and analysis of relative location of groups of three-dimensional objects (scene analysis).

6. Recognition of three-dimensional objects thrown together on top of one another.

[†]Specific recognition algorithms based on two-dimensional transformations are described in Vol. 3 of this series.

Tasks 1–3 are typical in, for instance, controlling robots that are performing assembly of products in a nonorderly arranged environment. Tasks 3–5 arise in the process of traffic control of an autonomous transportation robot. They are also frequently encountered by specialists implementing robots in poorly organized environments.

REFERENCES

Chapter 1

1. Andrew, A. Artificial intelligence. Moscow, Mir, 1985.

2. Efimov, E. I. Solvers of intelligent problems. Moscow, Nauka, 1982.

3. Kuzin, L. T. Fundamentals of cybernetics. Vol. 2. Fundamentals of cybernetic models. Moscow, Energiya, 1979.

4. Knowledge representation in man-machine and robotic systems. Vol. D. Fundamental and applied studies in the area of robotic systems. Moscow, VTs (Computer Center) AN SSSR, VINITI, 1984.

Chapter 2

1. Knowledge representation in man-machine and robotic systems. Vol. A. Fundamental studies in the area of knowledge representation. Moscow, VTs AN SSSR, VINITI, 1984.

2. Pupyrev, E. I. Adjustable automatic machines and microprocessor systems. Moscow, Nauka, 1984.

3. Novikov, P. S. Elements of mathematical logic. Moscow, Nauka, 1973.

4. Mendelson, E. Introduction to mathematical logic. Moscow, Nauka, 1984.

5. Popov, E. V. and Fridman, G. R. Algorithmic bases of intelligent robots and artificial intelligence. Moscow, Nauka, 1976.

6. Yatsuk, V. Ya. Utilization of λ-frames and a production method in developing intelligent systems of decision making. International symposium on artificial intelligence. Leningrad. October 4–6, 1983.

7. Knowledge representation in man-machine and robotic systems. Vol. B. Instruments for developing systems that utilize knowledge.

8. Pilshchikov, V. N. Language PLANNER. Moscow, Nauka, 1983.

Chapter 3

1. Nilsson, N. Artificial intelligence. Methods of solution search. Transl. from English. Moscow, Mir, 1973.

2. Besaker, R. and Saati, T. Finite graphs and networks. Transl. from English. Moscow, Mir, 1973.

3. Sleigl, J. Artificial intelligence. Moscow, Mir, 1973.

4. Hunt, E. Artificial intelligence. Moscow, Mir, 1978.

5. De Groot, M. Optimal statistical solutions. Moscow, Mir, 1974.

6. Zade, L. The concept of linguistic variable and its application to making approximate decisions. Moscow, Mir, 1976.

7. Bellman, R. and Zade, L. Decision making under diffuse conditions. In: Problems of analysis and procedures of decision making. Moscow, Mir, 1976.

Chapter 4

1. Pospelov, G. S. Artificial intelligence: the basis of new information technology. International symposium on Artificial Intelligence. Leningrad, October 4–6, 1983.

INDEX

This book is to be returned on or before
the last date stamped below.